LONDON MATHEMATICAL SOCIETY LECTURE NOTE SERIES

Managing Editor: Professor I.M.James,
Mathematical Institute, 24-29 St Giles, Oxford

1. General cohomology theory and K-theory, P.HILTON
4. Algebraic topology: a student's guide, J.F.ADAMS
5. Commutative algebra, J.T.KNIGHT
8. Integration and harmonic analysis on compact groups, R.E.EDWARDS
9. Elliptic functions and elliptic curves, P.DU VAL
10. Numerical ranges II, F.F.BONSALL & J.DUNCAN
11. New developments in topology, G.SEGAL (ed.)
12. Symposium on complex analysis, Canterbury, 1973, J.CLUNIE & W.K.HAYMAN (eds.)
13. Combinatorics: Proceedings of the British combinatorial conference 1973, T.P.McDONOUGH & V.C.MAVRON (eds.)
14. Analytic theory of abelian varieties, H.P.F.SWINNERTON-DYER
15. An introduction to topological groups, P.J.HIGGINS
16. Topics in finite groups, T.M.GAGEN
17. Differentiable germs and catastrophes, Th.BROCKER & L.LANDER
18. A geometric approach to homology theory, S.BUONCRISTIANO, C.P.ROURKE & B.J.SANDERSON
20. Sheaf theory, B.R.TENNISON
21. Automatic continuity of linear operators, A.M.SINCLAIR
23. Parallelisms of complete designs, P.J.CAMERON
24. The topology of Stiefel manifolds, I.M.JAMES
25. Lie groups and compact groups, J.F.PRICE
26. Transformation groups: Proceedings of the conference in the University of Newcastle upon Tyne, August 1976, C.KOSNIOWSKI
27. Skew field constructions, P.M.COHN
28. Brownian motion, Hardy spaces and bounded mean oscillation, K.E.PETERSEN
29. Pontryagin duality and the structure of locally compact abelian groups, S.A.MORRIS
30. Interaction models, N.L.BIGGS
31. Continuous crossed products and type III von Neumann algebras, A.VAN DAELE
32. Uniform algebras and Jensen measures, T.W.GAMELIN
33. Permutation groups and combinatorial structures, N.L.BIGGS & A.T.WHITE
34. Representation theory of Lie groups, M.F.ATIYAH et al.
35. Trace ideals and their applications, B.SIMON
36. Homological group theory, C.T.C.WALL (ed.)
37. Partially ordered rings and semi-algebraic geometry, G.W.BRUMFIEL
38. Surveys in combinatorics, B.BOLLOBAS (ed.)
39. Affine sets and affine groups, D.G.NORTHCOTT
40. Introduction to H_p spaces, P.J.KOOSIS
41. Theory and applications of Hopf bifurcation, B.D.HASSARD, N.D.KAZARINOFF & Y-H.WAN
42. Topics in the theory of group presentations, D.L.JOHNSON
43. Graphs, codes and designs, P.J.CAMERON & J.H.VAN LINT
44. Z/2-homotopy theory, M.C.CRABB
45. Recursion theory: its generalisations and applications, F.R.DRAKE & S.S.WAINER (eds.)
46. p-adic analysis: a short course on recent work, N.KOBLITZ
47. Coding the Universe, A. BELLER, R. JENSEN & P. WELCH
48. Low-dimensional topology, R. BROWN & T.L. THICKSTUN (eds.)
49. Finite geometries and designs, P. CAMERON, J.W.P. HIRSCHFELD & D.R. HUGHES (eds.)
50. Commutator Calculus and groups of homotopy classes, H.J. BAUES
51. Synthetic differential geometry, A. KOCK
52. Combinatorics, H.N.V. TEMPERLEY (ed.)
53. Singularity theory, V.I. ARNOLD
54. Markov processes and related problems of analysis, E.B. DYNKIN
55. Ordered permutation groups, A.M.W. GLASS
56. Journees arithmetiques 1980, J.V. ARMITAGE (ed.)
57. Techniques of geometric topology, R.A. FENN
58. Singularities of differentiable functions, J. MARTINET
59. Applicable differential geometry, F.A.E. PIRANI and M. CRAMPI
60. Integrable systems, S.P. NOVIKOV et al.

61. The core model, A. DODD
62. Economics for mathematicians, J.W.S. CASSELS
63. Continuous semigroups in Banach algebras, A.M. SINCLAIR
64. Basic concepts of enriched category theory, G.M. KELLY
65. Several complex variables and complex manifolds I, M.J. FIELD
66. Several complex variables and complex manifolds II, M.J. FIELD
67. Classification problems in ergodic theory, W. PARRY & S. TUNCEL

London Mathematical Society Lecture Note Series. 65

Several Complex Variables and Complex Manifolds I

MIKE FIELD
Senior Lecturer, Department of Pure Mathematics,
University of Sydney

CAMBRIDGE UNIVERSITY PRESS

CAMBRIDGE

LONDON NEW YORK NEW ROCHELLE

MELBOURNE SYDNEY

Published by the Press Syndicate of the University of Cambridge
The Pitt Building, Trumpington Street, Cambridge CB2 1RP
32 East 57th Street, New York, NY 10022, USA
296 Beaconsfield Parade, Middle Park, Melbourne 3206, Australia

First published in 1982

Printed in Great Britain at the University Press, Cambridge

Library of Congress catalogue card number 81-21590

British Library cataloguing in publication data

Field, Mike
Several complex variables and complex manifolds 1.-(London Mathematical Society Lecture Note Series 65, ISSN 0076-0552.)
1.Topological spaces 2. Manifolds (Mathematics)
I Title. II Series
514'.223 QA611.A1

ISBN 0 521 28301 9

Preface.

These notes, in two parts, are intended to provide a self-contained and relatively elementary introduction to functions of several complex variables and complex manifolds. They are based on courses on complex analysis that I have given at symposia at the International Centre for Theoretical Physics, Trieste, in 1972 and 1974 and various postgraduate and seminar courses held at Warwick and Sydney. Prerequisites for the reading of Part I are minimal and, in particular, I have made no significant use of differential forms, algebraic topology, differential geometry or sheaf theory. As these notes are primarily directed towards graduate and advanced undergraduate students I have included some exercises. There are also a number of references for further reading which may serve as a suitable starting point for graduate assignments or projects. I have endeavoured to give at least one reference for any result stated but not proved in the text. For the more experienced reader, who is not a specialist in complex analysis, I have included references to related topics not directly within the scope of these notes.

My aim in these notes was to give a broad introduction to several complex variables and complex manifolds and, in particular, achieve a synthesis of the theories of compact and non-compact complex manifolds. This approach is perhaps best exemplified by the conclusion of Part II where we present Grauert's pseudoconvexivity proof of the Kodaira embedding theorem. I would hope that parts I and II together comprise a useful introduction to more advanced works on complex analysis. Notably, the books by Grauert and Remmert on Stein spaces [1] and coherent analytic sheaves (forthcoming) and that of Griffiths and Harris on the Principles of Algebraic Geometry [1].

Chapter 1 of the text is devoted to functions of one complex variable and Riemann surfaces with particular emphasis on the $\bar{\partial}$-operator and the construction of meromorphic functions with specified pole and zero sets, themes that run throughout parts I and II. The presentation is geared towards generalisations to several variables and complex manifolds and most of the results, though perhaps not the proofs, should be familiar to all readers. Section 5 of the Chapter (on vector bundles) can safely be omitted on first reading. In Chapter 2, we develop the basic theory of analytic functions of several complex variables. Amongst the results and concepts discussed are Hartog's theorem on extension of analytic functions, domains of holomorphy, holomorphic convexivity, pseudoconvexivity, Levi pseudoconvexivity and the Levi problem, the Bergman kernel function, the Cousin problems. In section 5, I have given a fairly complete treatment of boundary invariants of domains in $\mathbb{R}^n$ with C^2 boundary. In part this was because of the incomplete treatment of the topic in other texts on several complex variables. In Chapter 3 we prove the Weierstrass Division and Preparation theorems and give applications to the algebraic structure of power series rings and the local structure theory of analytic sets. Here, as elsewhere in the notes, I have concentrated on the structure theory of hypersurfaces leaving the much harder general structure theory of analytic sets to the references (for example, Gunning-Rossi [1], Narasimhan [3], Whitney [1] and the forthcoming text by Grauert and Remmert on coherent analytic sheaves). The chapter concludes with a section on modules over power series rings, the reading of which may be deferred until Chapter 6 of Part II. In Chapter 4 we describe a number of basic examples of complex manifolds, both compact and non-compact, and conclude with sections on the structure theory of analytic hypersurfaces and blowing up.

Part II of these notes consists of three chapters which we now briefly describe. Chapter 5 covers calculus on complex manifolds including the construction of the $\bar{\partial}$-operator and the Dolbeault-Grothendieck lemma. Chapter 6 is a self-contained introduction to the theory of sheaves in complex analysis. Chapter 7 is devoted to coherence and the cohomology vanishing theorems of Cartan, Grauert and Serre. Applications include Grauert's proof of the Kodaira embedding theorem.

When I originally started these notes I had intended to include chapters on complex differential geometry and the elliptic theory of the complex Laplace-Beltrami operator applied to compact and non-compact complex manifolds. For reasons of length I eventually decided to omit these topics from Parts I and II. However, the reader will find references to chapters 8 through 12 scattered throughout the text. At some future time I hope it may be possible to complete the project with these additional chapters.

A few words of guidance to the reader of Part I: There is more than enough material in these notes for a one semester course. As we make the most substantial use of Chapter 3 in Part II, the reader may prefer to omit Chapter 3 at first reading, together with those parts of Chapter 4 on meromorphic functions and analytic sets (in particular, section 6). An alternative approach would be to read Chapter 3 (omitting section 6) and conclude with selected sections of Chapter 4 including section 6 on the structure theory of analytic hypersurfaces (this last section plays an important rôle in Part II).

I would like to acknowledge the great debt I owe in the preparation of these notes to many authors. I especially would like to mention the books by Grauert and Remmert on *Stein Spaces*, Gunning and Rossi on *Analytic functions of Several Complex Variables* and Hörmander on *Complex Analysis in Several Variables*. This last work has perhaps had the most decisive influence on the final form of my lecture notes.

On a more personal level, it is a great pleasure for me to express thanks to Jim Eells for interesting me in the field of complex analysis back in 1970 and for his continued help and encouragement since then. Thanks also to Tzee-Char Kuo for his advice and encouragement and to the postgraduate students here at Sydney who have been so helpful with their stimulating comments, assignments and critical questioning. Last, but by no means least, may I thank Cathy Kicinski for her beautiful job of typing the bulk of my manuscript.

Mike Field

Sydney,
September, 1981.

Contents

Notations and Conventions.

Throughout these notes $\mathbb{R}^n$ will always denote real n-space and $\mathbb{C}^n$ complex n-space. We shall often identify $\mathbb{C}^n$ and $\mathbb{R}^{2n}$ by letting $(z_1,\ldots,z_n) \in \mathbb{C}^n$ correspond to $(x_1,y_1,\ldots,x_n,y_n) \in \mathbb{R}^{2n}$, where $z_j = x_j + iy_j$, $1 \le j \le n$. We let $\mathbb{C}^\circ$ denote the multiplicative group of non-zero complex numbers. We let $\mathbb{Z}$, $\mathbb{N}$ denote the integers and positive integers respectively.

If E, F are finite dimensional vector spaces over the field $\mathbb{K}$, we let $L_{\mathbb{K}}(E,F)$ denote the $\mathbb{K}$-vector space of $\mathbb{K}$-linear maps from E to F. We often drop the subscript $\mathbb{K}$ when it is implicit from the context. If $\mathbb{K} = \mathbb{R}$, we set $E' = L_{\mathbb{R}}(E,\mathbb{R})$ and if $\mathbb{K} = \mathbb{C}$, we set $E^* = L_{\mathbb{C}}(E,\mathbb{C})$. If $A \in L_{\mathbb{R}}(E,F)$, we let $A' \in L_{\mathbb{R}}(F',E')$ denote the transpose of A. We let A^* denote the transpose of A in case A is $\mathbb{C}$-linear. We let GL(E) denote the group of linear isomorphisms of E. In case $E = \mathbb{R}^n$, we often use the notation $GL(n,\mathbb{R})$ for $GL(\mathbb{R}^n)$. Similarly, we often write $GL(n,\mathbb{C})$ instead of $GL(\mathbb{C}^n)$.

A *domain* will always refer to a connected open subset.

If Ω is a domain in $\mathbb{C}^n$, E is a finite dimensional vector space (over $\mathbb{R}$ or $\mathbb{C}$) and $f: \Omega \to E$ we say that f is C^r if it is r-times continuously differentiable. That is, we identify $\mathbb{C}^n$ with $\mathbb{R}^{2n}$ and require that all partial derivatives of f of order less than or equal to r exist and are continuous on Ω. If f is C^r for all positive integers r, we say that f is C^∞ (or smooth). We let $C^r(\Omega,E)$ denote the space of all E-valued C^r maps on Ω. In case $E = \mathbb{C}$ we abbreviate to $C^r(\Omega)$ and if $E = \mathbb{R}$, we abbreviate to $C^r_{\mathbb{R}}(\Omega)$.

If f is a vector valued map we define the (closed) support of f, supp(f), to be the closure of the set of points where f is non-zero. If supp(f) is compact we shall say that f has compact support. We denote the set of C^r E-valued maps on a domain Ω which have compact support by $C^r_c(\Omega,E)$.

Suppose that Ω is a domain in $\mathbb{C}^n$ or $\mathbb{R}^n$ and $f \in C^r(\Omega,E)$, $r > 0$. We use either of the notations D^sf_x, $D^sf(x)$ to denote the sth. derivative of f at x. Thus, D^sf_x will be an s-linear E-valued map (see also Dieudonné [1] and Field [1]).

Given $r > 0$, $z \in \mathbb{C}$, we let $D_r(z)$, $\bar{D}_r(z)$ denote the open and closed discs, centre z, radius r in $\mathbb{C}$ respectively. We let $D_r(z)^*$ denote the punctured disc $D_r(z) \setminus \{z\}$.

For $z = (z_1,\ldots,z_n) \in \mathbb{C}^n$ we define

$$|z| = \max_i |z_i|$$

$$\|z\| = \left\{\sum_{i=1}^{n} |z_i|^2\right\}^{\frac{1}{2}}, \text{ "Euclidean norm"}$$

For $r > 0$, we let $D(z;r)$, $E(z;r)$ respectively denote the open discs, centre z, radius r in $\mathbb{C}^n$ relative to norms $|\ |$, $\|\ \|$. Given, $r_1,\ldots,r_n > 0$, we let $D(z;r_1,\ldots,r_n)$ denote the open polydisc $\prod_{j=1}^{n} D_{r_j}(z_j) \subset \mathbb{C}^n$.

If f is a continuous $\mathbb{C}$- or $\mathbb{R}$-valued function defined on a neighbourhood of a compact set K, we define

$$\|f\|_K = \sup_{x\in K} |f(x)|.$$

Other notations will be defined in the text. We remark here only that from Chapter 5, $C^p(M)$ will refer to the space of smooth differential p-forms on the differential manifold M and that from Chapter 6, $\mathbb{C}$ may also refer to the constant sheaf of complex numbers.

Finally, we remark that Hermitian forms are always complex linear in the *first* variable.

CHAPTER 1. FUNCTIONS OF ONE COMPLEX VARIABLE

Introduction

Our aim in this chapter is to develop the familiar theories of analytic functions of one complex variable and Riemann surfaces in a way that generalises well to the several variable theory. In §3 we see how the existence theory of the Cauchy-Riemann equations can be used to prove the Mittag-Leffler theorem and also how the topology of domains in $\mathbb{C}$ naturally enters into the proof of the Weierstrass theorem. In §§4,5 we show how the theory of holomorphic line bundles may be used to reformulate some of the classical problems in Riemann surface theory. We also define the Cauchy-Riemann equations on an arbitrary Riemann surface and indicate how they are related to the problem of constructing meromorphic functions with specified divisors. In an appendix we prove a number of classical results, including the Runge approximation theorem. We use the Runge theorem to construct solutions of the Cauchy-Riemann equations.

§1. Analytic functions and power series

Let Ω be a domain in $\mathbb{C}$. We recall that a function $f: \Omega \to \mathbb{C}$ is said to be *analytic* or *holomorphic* if it is complex differentiable on Ω. Writing f in real and imaginary parts, $f = u + iv$, analyticity implies that u and v satisfy the Cauchy-Riemann equations on Ω:

$$\partial u/\partial x = \partial v/\partial y; \quad \partial u/\partial y = -\partial v/\partial x .$$

Recalling that a real 2×2-matrix $[a_{ij}]$ induces a complex linear endomorphism of $\mathbb{C}$ if and only if $a_{11} = a_{22}$ and $a_{12} = -a_{21}$, we may interpret the Cauchy-Riemann equations as saying that if f is analytic then f is differentiable in the real sense and the (real) derivative of f is everywhere a complex linear map (see, for example, Field [1; Example 3, page 133]).

Let $A(\Omega)$ denote the set of all analytic functions on Ω.

We now introduce a pair of partial differential operators which, together with their generalisations to several variables, will be of the utmost importance in the sequel. We set

$$\partial/\partial z = \tfrac{1}{2}(\partial/x - i\partial/\partial y); \ \partial/\partial\bar{z} = \tfrac{1}{2}(\partial/\partial x + i\partial/\partial y).$$

For $r \geq 1$,

$$\partial/\partial z, \partial/\partial\bar{z}: C^r(\Omega) \to C^{r-1}(\Omega).$$

The significance of these operators may be gauged from

Lemma 1.1.1. A function $f \in C^1(\Omega)$ is analytic if and only if $\partial f/\partial\bar{z} = 0$.

Proof. The reader may verify that $\partial f/\partial\bar{z} = 0$ iff the Cauchy-Riemann equations hold. Since f is assumed to be C^1, the Cauchy-Riemann equations hold iff f is analytic. □

Next we recall the basic theorem on the local representation of analytic functions by power series.

Theorem 1.1.2. Let $f \in A(\Omega)$. Given $\zeta \in \Omega$ and $r > 0$ such that $D_r(\zeta) \subset \Omega$, we have

$$f(z) = \sum_{j=0}^{\infty} a_j(z - \zeta)^j, \ z \in D_r(\zeta),$$

where $a_j = \partial^j f/\partial z^j(\zeta)/j!$, and convergence is uniform on compact subsets of $D_r(\zeta)$.

Remark. Simple examples, such as $f(z) = (1 - z)^{-1}$, $\Omega = \mathbb{C} \setminus \{1\}$, show that the power series at ζ need not converge on the whole of Ω.

Corollary 1.1.3. An analytic function is C^∞.

Remark. Notice that $A(\Omega) = \text{Kernel}(\partial/\partial\bar{z})$. Now $\partial/\partial\bar{z}$ is an example of an *elliptic* differential operator and it can be shown that the kernel of any elliptic operator consists of C^∞ functions. We shall return to this type of question in later chapters.

Corollary 1.1.4. If $f \in A(\Omega)$, then $\partial^j f/\partial z^j \in A(\Omega)$, $j \geq 0$.

Corollary 1.1.5. Let Ω be a domain in $\mathbb{C}$. Suppose $f \in A(\Omega)$ and that at some point $\zeta \in \Omega$, $\partial^j f/\partial z^j(\zeta) = 0$, $j \geq 0$. Then f vanishes identically on Ω.

Proof. Let $X = \{z \in \Omega: \partial^j f/\partial z^j = 0, \text{ all } j \geq 0\}$. X is open by the power series representation of analytic functions given by Theorem 1.1.2 and X is certainly non-empty since $\zeta \in X$. Since X is the intersection of the closed sets $\{z \in \Omega: \partial^j f/\partial z^j(z) = 0\}$, X is also closed. Since Ω is connected, $X = \Omega$. □

Remark. Another way of stating this corollary is that the value of an analytic function and all its derivatives at a single point of a domain determine the function uniquely. This type of behaviour does not, of course, hold for C^∞ or continuous functions.

Proposition 1.1.6. (Uniqueness of analytic continuation). Let U, V be connected open subsets of $\mathbb{C}$ and suppose $U \cap V \neq \emptyset$. If $f \in A(U)$ and h is an analytic extension of f to $U \cup V$ (that is, $h \in A(U \cup V)$ and $h|U = f$), then h is unique.

Proof. If h_1, h_2 are analytic extensions of f to $U \cup V$ then $h_1 - h_2$ is an analytic extension of the zero function on U to $U \cup V$. Therefore, $h_1 - h_2$ is identically zero by Corollary 1.1.5. □

Remark. Once we have uniqueness of analytic continuation it is natural to try to construct the largest domain to which any given analytic function may be extended. It turns out of course that we have to enlarge our class of domains to include Riemann surfaces spread over $\mathbb{C}$. We return to this question in §4 of this chapter and again in §2 of Chapter 6.

Exercises. These exercises are revision of basic theory of functions of one complex variable. Proofs may be found in any of the many introductory texts on complex analysis.

1) (Laurent series). Let f be analytic on the annulus $r < |z - z_0| < R$. Derive the Laurent series of f at z_0,

$$f(z) = \sum_{n=-\infty}^{n=+\infty} a_n(z - z_0)^n, \quad r < |z - z_0| < R,$$

where $a_n = (2\pi i)^{-1} \int_{|\zeta - z_0| = s} f(\zeta)/(\zeta - z_0)^{n+1} d\zeta$, $r < s < R$, and convergence is uniform on compact subsets of the annulus.

2) (Cauchy's inequalities). Continuing with the notation and assumptions of question 1, show that if $M(t) = \sup\{|f(z)|: |z - z_0| = t\}$, $r < t < R$, then

$$|a_n| \le M(t)/t^n, \; n \in \mathbb{Z} .$$

In particular, show that if f is holomorphic on the disc $D_R(z_0)$ then

$$|\partial^n f/\partial z^n(z_0)| \le M(t)n!/t^n, \; n \ge 0 .$$

3) (Riemann removable singularities theorem). Suppose that f is an analytic function in the punctured disc $D_r(z_0)^* = \{z: 0 < |z - z_0| < r\}$. Show that a necessary and sufficient condition for f to extend analytically to $D_r(z_0)$ is that f is bounded on some neighbourhood of z_0. (Use the result of question 2).

4) (Open mapping theorem). Let f be analytic and not identically zero on the domain U in $\mathbb{C}$. Prove that f(U) is open in $\mathbb{C}$.

5) (Monodromy theorem). Let Ω be a domain in $\mathbb{C}$, $z_0, y_0 \in \Omega$ and suppose f is analytic on some neighbourhood U of z_0. Let C be a continuous path in Ω parametrized by $\phi: [0,1] \to \Omega$ with $\phi(0) = z_0$, $\phi(1) = y_0$. We say f can be analytically continued along C if we can find discs $D_i = D_{r_i}(\phi(t_i))$, $0 = t_0 < \ldots < t_k = 1$, covering C, $h_i \in A(D_i)$, $i = 0,\ldots,k$, such that $h_0 = f$ on $U \cap D_0$ and $h_i = h_{i+1}$ on $D_i \cap D_{i+1}$, $i \ge 0$. Define $f_C(y_0)$ to be $h_k(y_0)$. Show that

a) $f_C(y_0)$ depends only on C and not on any of the choices we have made.

b) If C, C' are homotopic curves in Ω joining z_0 to y_0 then $f_C(y_0) = f_{C'}(y_0)$.

Give examples to show that if Ω is not simply connected and C, C' are non-homotopic curves joining z_0 to y_0 then $f_C(y_0)$ may not equal $f_{C'}(y_0)$.

6) (Maximum principle). Let f be analytic on the domain Ω in $\mathbb{C}$. If $|f|$ has a maximum in Ω then f is constant.

7) (Schwarz' lemma). Suppose $f \in A(D_1(0))$ and satisfies $f(0) = 0$ and $|f(z)| \le 1$, $z \in D_1(0)$. Then $|f(z)| \le z$ and $|f'(0)| \le 1$. Equality holds if and only if $f(z) = cz$, where $|c| = 1$ (Hint: Apply the maximum principle to the function $f(z)/z$).

§2. Meromorphic functions

Roughly speaking meromorphic functions are the analytic analogue of rational functions and in this section we briefly review their definition and elementary properties. Throughout the section Ω will denote a domain in $\mathbb{C}$.

Let $\zeta \in \Omega$ and $\Omega' = \Omega \setminus \{\zeta\}$. Suppose $f \in A(\Omega')$. By Laurent's expansion we have

$$f(z) = \sum_{j=-\infty}^{j=+\infty} a_j(z-\zeta)^j, \quad z \in D_r(\zeta)^* \subset \Omega.$$

There are three possibilities:

a) f is bounded on some neighbourhood of ζ. In this case $a_j = 0$, $j < 0$, and f extends uniquely to an analytic function on Ω (Riemann removable singularities theorem).

b) $f(z) \to \infty$ as $z \to \zeta$. Here one can show that there exists a strictly positive integer N such that $a_j = 0$, $j < -N$ and $a_{-N} \neq 0$.

c) Neither a) or b) occurs (ζ is an *essential singularity* of f).

In case b), f is an example of a *meromorphic function* on Ω with a single *pole* of order N at ζ. We may write $f(z) = u(z)/(z-\zeta)^N$, $z \in D_r(\zeta)^*$, where $u \in A(D_r(\zeta)^*)$ and the power series of u at ζ is given explicitly as the Laurent series of f at ζ multiplied by $(z-\zeta)^N$. We may extend u analytically to Ω by taking $u(z) = (z-\zeta)^N f(z)$, $z \neq \zeta$. In this way we may represent f as the quotient $u(z)/(z-\zeta)^N$ of analytic functions defined on all of Ω.

There are some problems in giving a satisfactory general definition of a meromorphic function. If we attempt to define a meromorphic function on Ω as a quotient f/g, $f,g \in A(\Omega)$, $g \not\equiv 0$, we are faced with the difficulty that f and g may have infinitely many common zeros. If this happens we canot cancel the zeros using the elementary power series techniques of the previous paragraph to obtain the maximal subdomain of Ω on which the meromorphic function is defined as an analytic function. That is, the representation f/g may be rather non-canonical. We start by giving a definition that is rather special to functions of one complex variable and then show how to reformulate the definition in a way that generalises well to functions of several complex variables.

Definition 1.2.1. We say that m is a *meromorphic* function on Ω if there exists a discrete subset X of Ω such that

i) m is an analytic function on $\Omega \setminus X$.

ii) Every point of X is a pole of m.

We notice that the definition excludes essential singularities and so, for example, $e^{-1/z}$ would not define a meromorphic function on $\mathbb{C}$.

We denote the set of meromorphic functions on Ω by $M(\Omega)$.

Locally a meromorphic function can be expressed as a quotient of analytic functions. That is, given $m \in M(\Omega)$ and $z \in \Omega$, we can find an open neighbourhood U of z and $f,g \in A(U)$ such that g is not identically zero and $m|U = f/g$ outside of any poles of m in U. Of course, if U does not contain any poles of m, we can take $g \equiv 1$.

We now work towards an alternative definition of a meromorphic function which is framed in terms of local information and requires no explicit information about the pole set.

Suppose that $\{U_i : i \in I\}$ is an open cover of Ω and that for each $i \in I$ we are given $f_i, g_i \in A(U_i)$ with g_i not vanishing identically on any connected component of U_i. Then $\{(f_i, g_i) : i \in I\}$ defines a meromorphic function m on Ω provided that for all $i,j \in I$ we have

$$f_i/g_i = f_j/g_j$$

at all points of $U_i \cap U_j$ where both g_i and g_j are non-zero. (Equivalently, $f_i g_j = f_j g_i$ on $U_i \cap U_j$). We omit the routine construction of m. If

$\{V_j: j \in J\}$ is another open cover of Ω and $\{(a_j,b_j): j \in J\}$ a corresponding set of analytic functions satisfying the above conditions, it is easily verified that $\{(f_i,g_i): i \in I\}$ and $\{(a_j,b_j): j \in J\}$ define the same meromorphic function if and only if

$$f_i/g_i = a_j/b_j$$

at all points of $U_i \cap V_j$ where both g_i and b_j are non-zero.

We can now use this condition to define an equivalence relation on all sets of pairs of analytic functions $\{(f_i,g_i): i \in I\}$ satisfying the requisite compatibility conditions. The equivalence classes of this relation are then defined to be meromorphic functions. This is essentially the approach that we adopt in later chapters.

One immediate consequence of our local description of meromorphic functions is that $M(\Omega)$ is a field (the connectedness of Ω is essential here to avoid zero divisors).

Suppose $m \in M(\Omega)$ and $\zeta \in \Omega$. We define the order of m at ζ, $\text{ord}(m,\zeta)$, to be the smallest index with non-zero coefficient in the Laurent expansion of m at ζ. That is, if

$$m(z) = \sum_{j=N}^{\infty} a_j (z-\zeta)^j$$

on some neighbourhood of ζ and $a_N \neq 0$, then $\text{ord}(m,\zeta) = N$. Clearly, if $m = f/g$ in some neighbourhood of ζ, $\text{ord}(m,\zeta) = \text{ord}(f,\zeta) - \text{ord}(g,\zeta)$ though of course the terms on the right hand side depend on the choice of local representation for m!

If $\text{ord}(m,\zeta) > 0$, we say that m has a *zero of order* $\text{ord}(m,\zeta)$ at ζ and if $\text{ord}(m,\zeta) < 0$, we say that m has a *pole of order* $-\text{ord}(m,\zeta)$ at ζ. We set

$$Z(m) = \{z \in \Omega: \text{ord}(m,z) > 0\} = \{z \in \Omega: m(z) = 0\}$$

$$P(m) = \{z \in \Omega: \text{ord}(m,z) < 0\} = \{z \in \Omega: m \text{ has a pole at } z\}.$$

$Z(m)$ and $P(m)$ are called the zero and pole set of m respectively. Clearly $Z(m)$ and $P(m)$ are disjoint discrete subsets of Ω (assuming $m \neq 0$).

We now introduce some useful definitions and notation. Suppose $p: \Omega \to \mathbb{Z}$ and $\{z \in \Omega: p(z) \neq 0\}$ is a discrete subset of Ω. We call the formal sum $\sum_{z\in\Omega} p(z).z$ a *divisor* on Ω. We denote the set of divisors on Ω by $\mathcal{D}(\Omega)$. $\mathcal{D}(\Omega)$ has the structure of an ordered abelian group if we define

$$\left(\sum_{z\in\Omega} p(z).z \pm \sum_{z\in\Omega} q(z).z\right) = \sum_{z\in\Omega} (p \pm q)(z).z$$

$$\sum_{z\in\Omega} p(z).z > \sum_{z\in\Omega} q(z).z \quad \text{if and only if}$$

$p(z) \geq q(z)$ for all $z \in \Omega$ with strict inequality for at least one point of Ω.

Let $M^*(\Omega)$ denote the group of invertible elements of $M(\Omega)$. Since Ω is assumed connected, $M^*(\Omega)$ is all of $M(\Omega)$ except the zero function. Given $m \in M^*(\Omega)$, the divisor of m, div(m), is defined to be

$$\sum_{z\in\Omega} \text{ord}(m,z).z$$

div: $M^*(\Omega) \to \mathcal{D}(\Omega)$ is a homomorphism (relative to the multiplicative structure on $M^*(\Omega)$). We note that $\text{div}(m) \geq 0$ if and only if $m \in A(\Omega)$.

Suppose $m \in M(\Omega)$ and $\zeta \in \Omega$. Let m have Laurent series

$$\sum_{j=N}^{\infty} a_j(z-\zeta)^j$$

at ζ, where we suppose $a_N \neq 0$. The principal part of m at ζ is defined to be

$$\sum_{j=N}^{-1} a_j(z-\zeta)^j$$

if $N \leq -1$ and zero otherwise. We note that $m, m' \in M(\Omega)$ have the same principal part at ζ if and only if $m - m'$ is analytic on some neighbourhood of ζ.

To conclude this section we remark that Proposition 1.1.6 generalises to meromorphic functions and therefore we can discuss meromorphic continuation. We leave details to the reader.

Exercises

1. Verify that

$$\text{div}(mm') = \text{div}(m) + \text{div}(m'),\ m,m' \in M^*(\Omega).$$

$$\text{div}(m^{-1}) = -\text{div}(m),\ m \in M^*(\Omega).$$

$\text{div}(m) = 0$ if and only if m is a nowhere vanishing analytic function on Ω.

2. Let $m \in M^*(\mathbb{C})$. Show that if $\text{div}(m) = 0$ then either m is constant or m has an essential singularity at infinity.

3. Under what conditions is the composition of two meromorphic functions metomorphic?

§3. Theorems of Weierstrass and Mittag-Leffler

In the preceding sections we have reviewed some of the basic elementary properties of analytic and meromorphic functions. However, we have as yet given no means of constructing such functions so as to satisfy specified properties. For example, if X is any closed subset of $\mathbb{C}$ it is not difficult to construct a C^∞ function on $\mathbb{C}$ with zero set X. Can we find an analytic function whose zero set is equal to X? Clearly we cannot unless X is a discrete subset of $\mathbb{C}$. It turns out though that if X is discrete we can always find an analytic function on $\mathbb{C}$ with zero set X. This is exactly the type of result we need if our study of analytic functions is to amount to much more than a study of polynomials, rational functions and the standard analytic functions such as log and exp.

Our aim in this section will be to show the importance of the theory of the partial differential operator $\partial/\partial\bar{z}$ and the topology of domains in $\mathbb{C}$ in questions involving the construction of analytic and meromorphic functions with specified behaviour at prescribed poles and zeros. We adopt this approach because it generalises well to functions of several complex variables and complex manifolds. We must emphasise, however, that the Mittag-Leffler and Weierstrass theorems can be given rather more elementary proofs than those presented here which do not depend on the theory of $\partial/\partial\bar{z}$ (see, for example, Heins [1] or Hille [1]).

Throughout this section Ω will denote an open subset of $\mathbb{C}$.

We give the proof of the following basic existence theorem in the appendix to this chapter.

Theorem 1.3.1. Let $f \in C^\infty(\Omega)$. Then there exists $u \in C^\infty(\Omega)$ such that

$$\partial u/\partial\bar{z} = f .$$

Remark. An equivalent formulation of Theorem 1.3.1 is that the sequence

$$0 \to A(\Omega) \xrightarrow{i} C^\infty(\Omega) \xrightarrow{\partial/\partial\bar{z}} C^\infty(\Omega) \to 0$$

is exact for every open subset Ω of $\mathbb{C}$ (i denotes inclusion).

The Mittag-Leffler theorem gives conditions under which there exists a meromorphic function on Ω with specified principal parts.

Before stating the Mittag-Leffler theorem we introduce some useful notation. Suppose U_i, U_j are sets, then we let U_{ij} denote the intersection $U_i \cap U_j$. We use the obvious generalisation of this notation for intersections of more than two indexed sets.

Theorem 1.3.2. (Mittag-Leffler theorem) Let $\{U_i: i \in I\}$ be an open cover of Ω and suppose we are given $m_i \in M(U_i)$ for each $i \in I$. Then, provided that $m_i - m_j \in A(U_{ij})$ for all $i,j \in I$, there exists $m \in M(\Omega)$ such that

$$m - m_i \in A(U_i),\ i \in I .$$

Remark. An alternative formulation of this theorem would be: Suppose X is a discrete subset of Ω and that for each $z \in X$ we are given a meromorphic function m^z which is defined on some neighbourhood of z and has a single pole at z. Then there exists $m \in M(\Omega)$ with pole set X and such that the principal part of m at z equals the principal part of m^z at z for all $z \in X$.

Proof of theorem. (Following Hörmander [1]). Observe that it suffices to construct $f_i \in A(U_i)$ such that for all i,j

$$m_i + f_i = m_j + f_j \text{ on } U_{ij}.$$

Indeed, if these compatibility conditions hold we can define $m \in M(\Omega)$ by setting $m|U_i = m_i + f_i$ and m will then obviously satisfy the requirements of the theorem.

What we shall do is to start by constructing $h_i \in C^\infty(U_1)$ such that $m_i + h_i = m_j + h_j$ on U_{ij}.This will only use the theory of C^∞ functions and no complex analysis. Using Theorem 1.3.1, we then construct a "correction" term $u \in C^\infty(\Omega)$ such that for all $i \in I$, $h_i + u \in A(U_i)$. It is then enough to take $f_i = h_i + u$.

Step 1. Choose a partition of unity $\{\theta_\alpha: \alpha \in \Lambda\}$ subordinate to the cover $\{U_i: i \in I\}$. That is, each θ_α is a positive C^∞ function with compact support, $\text{supp}(\theta_\alpha) \subseteq U_{\tau(\alpha)}$ for some $\tau(\alpha) \in I$, only finitely many θ_α are non-zero on any given compact subset of Ω and $\sum_{\alpha\in\Lambda} \theta_\alpha \equiv 1$ on Ω (for the construction of partitions of unity we refer to Lang [1], Spivak [1] or de Rham [1]).

Set $f_{ij} = m_i - m_j \in A(U_{ij})$ and observe that for all $i,j,k \in I$

$$f_{ij} + f_{jk} + f_{ki} = 0 \text{ on } U_{ijk} \text{ and } f_{ij} = -f_{ji} \text{ on } U_{ij} \qquad \text{...(A)}$$

We define

$$h_i = \sum_{\alpha\in\Lambda} \theta_\alpha f_{\tau(\alpha)i}.$$

Setting $f_{\tau(\alpha)i} = 0$ outside $U_{\tau(\alpha)i}$ it is clear that $h_i \in C^\infty(U_i)$. Moreover

$$\begin{aligned} h_j - h_i &= \sum_{\alpha\in\Lambda} \theta_\alpha (f_{\tau(\alpha)j} - f_{\tau(\alpha)i}) \\ &= \sum_{\alpha\in\Lambda} \theta_\alpha f_{ij}, \text{ by (A)} \\ &= f_{ij} = m_i - m_j. \end{aligned}$$

Hence we have found $h_i \in C^\infty(U_i)$ satisfying $m_i + h_i = m_j + h_j$ on U_{ij}.

Step 2. Since $h_i - h_j \in A(U_{ij})$,

$$\partial h_i/\partial \bar{z} = \partial h_j/\partial \bar{z} \text{ on } U_{ij} .$$

Hence we may define $F \in C^\infty(\Omega)$ by requiring that $F|U_i = \partial h_i/\partial \bar{z}$.

To construct $u \in C^\infty(\Omega)$ satisfying $u + h_i \in A(U_i)$ for all $i \in I$, we require $\partial u/\partial \bar{z} + \partial h_i/\partial \bar{z} = 0$ on U_i. That is,

$$\partial u/\partial \bar{z} = -F \text{ on } \Omega.$$

By Theorem 1.3.1, there exists $u \in C^\infty(\Omega)$ satisfying this equation. Finally we take $f_i = u + h_i \in A(U_i)$ and define m as indicated in the introduction to the proof. □

The relation (A) occurring in the proof of Theorem 1.3.2 is usually referred to as a *cocycle* condition and $\{f_{ij}: i,j \in I\}$ as a cocycle. We shall meet this type of relation frequently in the sequel. The proof of Theorem 1.3.2 clearly proves the slightly stronger

Theorem 1.3.2'. Let $f_{ij} \in A(U_{ij})$ satisfy the cocycle conditions

$$f_{ij} + f_{jk} + f_{ki} = 0 \text{ on } U_{ijk} \text{ and } f_{ij} = -f_{ji} \text{ on } U_{ij}, \text{ all } i,j,k.$$

Then there exist $g_i \in A(U_i)$ such that for all i,j,

$$f_{ij} = g_j - g_i \text{ on } U_{ij} .$$

<u>Notation</u>. If U is an open subset of $\mathbb{C}$, we shall let $A^*(U)$ denote the set of nowhere vanishing analytic functions on U. That is, the units in $A(U)$.

Theorem 1.3.3. (Weierstrass theorem) Let $\mathcal{U} = \{U_i: i \in I\}$ be an open cover of Ω and suppose that we are given $m_i \in M^*(U_i)$ for each $i \in I$. Then, provided that $m_i/m_j \in A^*(U_{ij})$ for all $i,j \in I$, there exists $m \in M^*(\Omega)$ such that

$$m/m_i \in A^*(U_i), \text{ for all } i \in I.$$

Equivalently: The map div: $M^*(\Omega) \to \mathcal{D}(\Omega)$ is surjective. That is, subject to the requirement that pole and zero sets are discrete, we may construct a meromorphic function on Ω with prescribed zeros and poles of given orders.

Proof. By taking a refinement of the cover $\mathcal{U}$, it is no loss of generality to assume that each $U_i \in \mathcal{U}$ is convex (for example, an open disc).

Given $i,j \in I$, set $h_{ij} = m_i/m_j$. Fix a branch of $\log h_{ij}$ on U_{ij} and define

$$c_{ijk} = \frac{1}{2\pi i}(\log h_{ij} + \log h_{jk} + \log h_{ki}).$$

Since $h_{ij}h_{jk}h_{ki} \equiv 1$, $c_{ijk} \in \mathbb{Z}$ for all i,j,k (note that U_{ijk} is convex and therefore connected and so c_{ijk} is constant on U_{ijk}).

Suppose that we can choose the branches of $\log h_{ij}$ so that $c_{ijk} = 0$ for all $i,j,k \in I$. Then if we set $f_{ij} = \log h_{ij} \in A(U_{ij})$, we see that the conditions of Theorem 1.3.2' are satisfied. Hence there exist $g_i \in A(U_i)$ such that for all $i,j \in I$

$$\log h_{ij} = g_j - g_i \text{ on } U_{ij} .$$

Now define $a_i = \exp(g_i) \in A^*(U_i)$. Clearly $a_j/a_i = h_{ij} = m_i/m_j$ and so

$$m_i a_i = m_j a_j \text{ on } U_{ij} .$$

Hence we can define $m \in M^*(\Omega)$ by taking $m|U_i = m_i a_i$. Clearly m satisfies the required conditions.

To complete the proof it is sufficient to show that we can choose the branches of $\log h_{ij}$ so that $c_{ijk} = 0$ for all i,j,k. This is essentially a topological problem. First we observe that $\{c_{ijk}\}$ defines a class in $H^2(\mathcal{U},\mathbb{Z})$ (For those unfamiliar with Čech theory we refer to Chapter 6). Now any finite intersection U of open sets of the cover $\mathcal{U}$ is convex and so $H^p(U,\mathbb{Z}) = 0$, $p \neq 0$. Hence by Leray's theorem (Theorem 6.3.16)

$$H^2(\mathcal{U},\mathbb{Z}) \cong H^2(\Omega,\mathbb{Z}) .$$

But $H^2(\Omega,\mathbb{Z}) = 0$ for all open domains in $\mathbb{C}$ (in fact for all non-compact oriented 2-manifolds - see the remarks following this proof). Hence $\{c_{ijk}\}$ is a coboundary and there exist integers n_{ij}, $i,j \in I$, such that

$$c_{ijk} = n_{ij} + n_{jk} + n_{ki}, \text{ for all } i,j,k \in I.$$

Now define a new branch of $\log h_{ij}$ by subtracting $2\pi i n_{ij}$ from the original choice. For the new choice of branches we have $c_{ijk} = 0$, for all i,j,k. □

Notice that the proof of Theorem 1.3.3 actually proves the slightly stronger

Theorem 1.3.3'. Let $h_{ij} \in A^*(U_{ij})$ satisfy the cocycle conditions

$$h_{ij}h_{jk}h_{ki} = 1 \text{ on } U_{ijk} \text{ and } h_{ij} = h_{ji}^{-1} \text{ on } U_{ij},\ i,j,k \in I.$$

Then there exist $h_i \in A^*(U_i)$ such that for all i,j

$$h_{ij} = h_j h_i^{-1} \text{ on } U_{ij} .$$

Remarks

1. The reader should observe the similarity between the proofs of Theorems 1.3.2 and 1.3.3. Indeed the last part of the above proof can be written as a problem in C^∞ functions if we note that $c_{ijk} = 0$ for all i,j,k iff there exists $b_i \in C^\infty(U_i)$ such that $\log h_{ij} = b_j/b_i$. Later, in Chapter 6, we develop machinery that handles all arguments of this type in a particularly simple and elegant way.

2. The explicit use of cohomology can of course be avoided in the proof of Weierstrass' theorem. See, for example, the proof given in Hörmander [1]. The reader might wish to attempt a direct construction of the integers n_{ij} without using any general facts about the cohomology of 2-manifolds.

3. For the vanishing of $H^2(\Omega,\mathbb{Z})$, we note that $H^2(\Omega,\mathbb{Z}) \cong H_2(\Omega,\mathbb{Z})$ (computation using universal coefficient theorems and the fact

that Ω is a 2-manifold). Since Ω is assumed oriented, Poincaré duality implies that $H_2(\Omega,\mathbb{Z}) \cong H_c^0(\Omega,\mathbb{Z})$, where c denotes compact supports. Since Ω is non-compact, $H_c^0(\Omega,\mathbb{Z}) = 0$. For further details and references we refer the reader to the appendix in Milnor and Stasheff [1].

Corollary 1.3.4. Let $m \in M(\Omega)$. Then there exist $f,g \in A(\Omega)$ such that

i) Off the pole set of m, $m = f/g$.

ii) $Z(m) = Z(f)$, $P(m) = Z(g)$ and $Z(f) \cap Z(g) = \emptyset$.

Proof. Suppose m has poles z_j with orders p_j. Then, by Weierstrass' theorem there exists $g \in A(\Omega)$ such that $\operatorname{div}(g) = \sum_j p_j.z_j$. Since $\operatorname{div}(mg) \geq 0$, $mg \in A(\Omega)$. Taking $f = mg$, it is clear that f and g satisfy the conditions of the corollary. □

Remark. f and g are unique up to multiplication by elements of $A^*(\Omega)$.

Corollary 1.3.5. There exists $f \in A(\Omega)$ which cannot be extended as an analytic or meromorphic function to any open subset strictly containing Ω.

Proof. Let $d(z,\partial\Omega)$ denote the distance between $z \in \mathbb{C}$ and the boundary of Ω, $\partial\Omega$. Define

$$X = \left\{\left(\frac{p}{2^n},\frac{q}{2^n}\right) \in \Omega : p,q,n \in \mathbb{Z},\ n \geq 0 \text{ and } d\left(\left(\frac{p}{2^n},\frac{q}{2^n}\right),\partial\Omega\right) < 2^{-n+2}\right\}.$$

X is a discrete subset of Ω and every point of $\partial\Omega$ is the limit of some subsequence of X. By Weierstrass' theorem there exists $f \in A(\Omega)$ such that

$$\operatorname{div}(f) = \sum_{x\in X} 1.x\ .$$

Since the zeros of an analytic or meromorphic function are isolated, we see that f cannot be extended to any domain strictly containing Ω. Indeed, such a domain would contain at least one point of $\partial\Omega$ and this point would then be a non-isolated zero of the extension. □

Remark. The proof of Corollary 1.3.5 actually shows that we can find an analytic function on Ω which cannot be continued (locally) across any point of the boundary of Ω.

Exercise. Let $\{z_i: i \geq 0\}$ be a discrete subset of $\Omega \subset \mathbb{C}$ and suppose that for each i we are given a polynomial $P_i(z) = \sum_{j=0}^{k_i} a_{ij}(z - z_i)^j$. Show that there exists an analytic function f on Ω such that the Taylor series of f at each z_i agrees with P_i to order k_i. (Hint: Find a meromorphic function m on Ω such that at each z_i the principal part of m equals $(z - z_i)^{-k_i-1} P_i(z)$. Then use Weierstrass' theorem).

§4. Riemann surfaces

In sections 4 and 5 we make a preliminary investigation of possible generalisations of the results of §3 to Riemann surfaces. As we shall see both the topology and complex structure of a Riemann surface play a crucial role as is also the case in the theory of higher dimensional complex manifolds.

Let U and V be open subsets of $\mathbb{C}$ and $f: U \to V$. We shall say that f is *biholomorphic* if f is a homeomorphism onto V and both f and f^{-1} are analytic. We remark that, by the inverse function theorem, for f to be biholomorphic it is sufficient for f to be bijective and everywhere analytic with non-vanishing derivative.

Before recalling the definition of a Riemann surface, we wish to stress that all topological spaces considered in this book will be Hausdorff and paracompact. Unless the contrary is clearly indicated they will also be connected. The Hausdorff and paracompactness assumptions imply metrizability and, together with connectedness, imply separability (see, for example, Matsushima [1]).

A *chart* on a topological space M consists of a pair (U,ϕ) where U is an open subset of M and ϕ is a homeomorphism of U onto an open subset of Euclidean space.

Definition 1.4.1. Let M be a topological space. Suppose that we are given a set $A = \{(U_i,\phi_i): i \in I\}$ of charts on M satisfying

1. $\{U_i\}$ is an open cover of M.

2. For each $i \in I$, ϕ_i is a homeomorphism of U_i onto an open subset of $\mathbb{C}$.

3. For all $i,j \in I$, $\phi_i\phi_j^{-1}$ is a biholomorphism of $\phi_j(U_{ij})$ with $\phi_i(U_{ij})$.

Then A is called a (*complex analytic*) *atlas* on M and M, together with the atlas A, is called a *Riemann surface*.

Remarks.

1. We refer to charts $(U,\phi) \in A$ as complex analytic charts on M or just charts on M if the analyticity is clear from the context.

2. If A is an atlas on M, we shall generally assume that A is maximal in the sense that if (U,ϕ) is a chart on M such that $\phi\phi_i^{-1}$ is biholomorphic for all $(U_i,\phi_i) \in A$ then $(U,\phi) \in A$. Of course, every atlas can be completed to a maximal atlas.

Before giving examples and motivation for the study of Riemann surfaces we shall generalise some of the definitions of §§1,2.

Suppose that M is a Riemann surface with atlas A. Let W be an open subset of M. If $f: W \to \mathbb{C}$, we say f is analytic if $f\phi^{-1} \in A(\phi(U \cap W))$ for all $(U,\phi) \in A$. We let $A(W)$ denote the ring of analytic functions on W. We say that m is a meromorphic function on W if $m\phi^{-1} \in M(\phi(U \cap W))$ for all $(U,\phi) \in A$ (a more careful definition can be made along the lines of Definition 1.2.1). We let $M(W)$ denote the ring of meromorphic functions on W and $M^*(W)$, $A^*(W)$ denote the groups of units in $M(W)$, $A(W)$ respectively. Note that if W is connected $M(W)$ is a field.

Let $m \in M^*(M)$. We define the zero set $Z(m)$ of m by

$$Z(m) = \cup \phi^{-1} Z(m\phi^{-1}) ,$$

where the union is over all charts $(U,\phi) \in A$. We similarly define the pole set $P(m)$ of m. We define the *order of m at z*, $\mathrm{ord}(m,z)$ to be $\mathrm{ord}(m\phi^{-1},\phi(z))$, where $(U,\phi) \in A$ contains z, and the *divisor* of m, $\mathrm{div}(m)$, by

$$\operatorname{div}(m) = \sum_{z \in M} \operatorname{ord}(m,z).z$$

We leave it to the reader to verify that $Z(m)$ and $P(m)$ are well-defined, discrete, disjoint subsets of M; that $M \setminus P(m)$ is the largest subset of M on which m is analytic; that ord(m,z) does not depend on the choice of chart containing z and that if M is compact, $Z(m)$ and $P(m)$ are finite.

Suppose N is another Riemann surface with atlas $\mathcal{B}$ and $f: M \to N$. We say that f is analytic or holomorphic if $\psi f \phi^{-1}: \phi(U \cap f^{-1}(V)) \to \mathbb{C}$ is analytic for all $(U,\phi) \in \mathcal{A}$, $(V,\psi) \in \mathcal{B}$ and that f is biholomorphic if f is bijective and both f and f^{-1} are analytic. If there exists a bilhomorphic map between M and N, we say that M and N are biholomorphically equivalent or analytically isomorphic.

Examples.

1. The *Riemann sphere*. The Riemann sphere is defined to be the 1-point compactification of $\mathbb{C}$, $\mathbb{C} \cup \{\infty\}$, and we shall denote it by either S^2 or $P^1(\mathbb{C})$ (see Chapter 4 for the second notation). We define an atlas $\{(U,\phi),(V,\psi)\}$ on S^2 by taking $U = \mathbb{C}$, $V = \mathbb{C}^{\bullet} \cup \{\infty\}$($\mathbb{C}^{\bullet}$ here denotes non-zero complex numbers) and $\phi(z) = z$: $\psi(z) = 1/z$, $\psi(\infty) = 0$.

Suppose Ω is an open subset of $\mathbb{C}$ and $m \in M(\Omega)$. Then m induces an analytic map $\tilde{m}: \Omega \to S^2$ if we take $\tilde{m}|\Omega \setminus P(m) = m$ and set $\tilde{m}(p) = \infty$, for all $p \in P(m)$. The verification that $\tilde{m}$ is analytic is easy using our explicit atlas for S^2.

2. If M is a *simply connected* Riemann surface then M is biholomorphic to precisely one of: The Riemann sphere, the open unit disc, the complex plane. This is the fundamental (and difficult!) theorem of Riemann surfaces known as the *Uniformization Theorem* (for a proof and more details we refer to Ahlfors and Sario [1], Siegel [1], Springer [1]).

3. Let $f: M \to N$ be a local homeomorphism between topological spaces M and N. That is, for each $x \in M$ there exists an open neighbourhood U of x such that f(U) is an open subset of N and $f|U$ maps U homeomorphically onto f(U). Suppose that N is a Riemann surface with atlas $\mathcal{A}$. Then f induces the structure of a Riemann surface on M in such a way that f becomes a holomorphic map which is locally biholomorphic. Indeed, suppose U is an open subset of M which is mapped homeomorphically by f onto an

open subset of N. Suppose $(W,\phi) \in A$ and $W \supset f(U)$. Then $(U,\phi f)$ is a chart on M. The set of all charts on M constructed in this way is easily seen to define a complex analytic atlas on M with respect to which f is locally biholomorphic.

4. By examples 2 and 3, the simply connected covering surface of a Riemann surface has the natural structure of a Riemann surface and is biholomorphic to one of: the Riemann sphere, the open unit disc, the complex plane (see also Chapter 4).

5. If M is a compact Riemann surface, M is diffeomorphic to a compact orientable surface. Such surfaces are classified, up to diffeomorphism by their *genus* or number of handles (see Hirsch [1]). However, genus does not classify M up to biholomorphic equivalence (unless M is simply connected). For a specific example, we refer to Chapter 4, §4.

6. Suppose $\pi: M \to \mathbb{C}$ is a local homeomorphism. Then M has the structure of a non-compact Riemann surface with respect to which π is locally biholomorphic. The pair (M,π) is called a *Riemann domain* or *spreading* of M over $\mathbb{C}$. Riemann domains occur as the domains of maximal analytic continuation of analytic functions defined on open subsets of $\mathbb{C}$. Indeed, if Ω is a domain in $\mathbb{C}$ and $f \in A(\Omega)$, it is not hard to construct a Riemann domain (M_f,π) such that f extends analytically to M_f and (M_f,π) is the maximal Riemann domain to which f can be continued analytically (The converse is also true but much harder: Any Riemann domain is the domain of maximal analytic continuation of an analytic function. This result, due to Behnke and Stein [1], generalises Corollary 1.3.5 of the Weierstrass theorem. See also R. Narasimhan [1] and Gunning and Rossi [1]). We give one explicit example here and leave the general construction to Chapter 6.

Let Log denote the principal branch of the logarithm defined off the negative real axis. We let $M \subset \mathbb{C}^2$ denote the set of all points $(z,\log|z| + i\arg z)$, $z \in \mathbb{C}^\bullet$, where arg z is a value of the argument of z. If we let π denote the restriction to M of the projection of $\mathbb{C}^2$ on the first factor, we may easily verify that (M,π) is a Riemann domain. Furthermore, M is the maximal Riemann domain to which Log extends analytically. Moreover, if we let $\log: M \to \mathbb{C}$ denote the restriction to M of the projection of $\mathbb{C}^2$ on the second factor, we see that log is the analytic extension of Log to M and that if $y \in \pi^{-1}(z)$, $z \in \mathbb{C}^{\bullet}$, then $\log(y) = \log|z| + i\phi$, where ϕ is a value of arg z.

7. There is no natural way of compactifying the Riemann domain associated to the logarithm that we constructed in the previous example. This is a reflection of the transcendental (non-algebraic) character of log (see Chapter 7). In the case of algebraic functions we can always compactify the Riemann domain associated to the function in a particularly nice way. Here we shall consider a simple example and refer the reader to Ahlfors [1] or Siegel [1] for the general theory.

Let us regard $P(y,z) = y^3 - z^2$ as a polynomial in y with coefficients depending on z. The zero set of P defines a subset X of $\mathbb{C}^2$. Now X will not be a complex submanifold of $\mathbb{C}^2$ as there is a *singularity* or *branch point* at $z = 0$ (see Chapter 4 for terminology). Let $\pi: X \to \mathbb{C}$ denote the restriction to X of the projection of $\mathbb{C}^2$ on the z-axis. If X_P denotes the 1-point compactification of X then, if we define $\pi(\infty) = \infty$, π extends continuously to X_P and $\pi: X_P \to S^2$ is a 3-fold branched cover of S^2 with branch points at 0 and ∞. In particular, if we set $X^* = X \setminus \{0,\infty\}$, $\pi: X^* \to \mathbb{C}^{\bullet}$ is a local homeomorphism and $\pi^{-1}(z)$ contains precisely three points for all $z \in \mathbb{C}^{\bullet}$. Now, by example 3, (X^*,π) is a Riemann domain and we denote its atlas by A^*. We now construct a complex analytic atlas for the whole of X_P with respect to which π is analytic. Let D denote the open unit disc centre zero in the ζ-plane. We define

$$\phi: D \to X_P \text{ by } \phi(\zeta) = (\zeta^2,\zeta^3)$$

$$\psi: D \to X_P \text{ by } \phi(\zeta) = (\zeta^{-2},\zeta^{-3}),\ \zeta \neq 0$$

$$= \infty,\ \zeta = 0.$$

Notice that ϕ and ψ are homeomorphisms onto open neighbourhoods of $\pi^{-1}(0)$ and ∞ respectively. As atlas on X_P we take

$$\{(\phi(D),\phi^{-1}),\ (\psi(D),\psi^{-1})\} \cup A^* .$$

We leave it to the reader to verify that this atlas defines on X_P the structure of a compact Riemann surface, biholomorphic to the Riemann sphere, with respect to which π is analytic.

Let $p: X_p \to S^2$ denote the restriction to X_p of the projection of $\mathbb{C}^2$ onto the second factor where we have defined $p(\infty) = \infty$. Then p is a meromorphic function on X_p with a pole of order 3 at ∞ and a zero of order 3 at 0. p gives all the roots of $y^3 = z^2$ in the sense that $\{p(x): x \in \pi^{-1}(z)\}$ is the set of all roots of $y^3 = z^2$ for all $z \in \mathbb{C}$.

For the remainder of this section we wish to investigate the possible generalisation of Weierstrass' theorem to compact Riemann surfaces. So from now on assume M is a compact Riemann surface.

Lemma 1.4.2. Every analytic function on M is constant.

Proof. If $f \in A(M)$, $|f|$ has a maximum on M, say at z_0. A simple application of the maximum modulus theorem shows that f is constant in some neighbourhood of z_0. Uniqueness of analytic continuation then implies that f is constant on M. □

In view of this Lemma it becomes a matter of some importance to prove that there exist plenty of non-constant meromorphic functions on M. Specifically: Let $d \in \mathcal{D}(M)$. Under what conditions on d, if any, do there exist $m \in \mathcal{M}^*(M)$ such that $\operatorname{div}(m) = d$?

For the Riemann sphere this question is easily solved:
If $d = \sum_{i=1}^{n} p_i . z_i \in \mathcal{D}(S^2)$, then there exists $m \in \mathcal{M}^*(S^2)$ with $\operatorname{div}(m) = d$ if and only if $\sum_{i=1}^{n} p_i = 0$.

Indeed, if $z_1, \ldots, z_{n-1} \in \mathbb{C}$, $z_n = \infty$, we may define

$$m(z) = \prod_{i=1}^{n-1} (z - z_i)^{p_i}, \quad z \in \mathbb{C} .$$

m extends to a meromorphic function on S^2 with singularities of order p_i at z_i, $1 \le i \le n-1$, and of order $-\sum_{i=1}^{n-1} p_i$ at ∞.

The converse is included in Lemma 1.4.3. below.

In future if $d = \sum_{z \in M} p(z).z$ is a divisor on M, we set

$$\deg(d) = \sum_{z \in M} p(z) \in \mathbb{Z}$$

and call deg(d) the *degree* of d.

We have the following topological restriction on divisors of meromorphic functions.

Lemma 1.4.3. A necessary condition for $d \in \mathcal{D}(M)$ to be the divisor of a meromorphic function is that $\deg(d) = 0$.

Proof. Let $m \in \mathcal{M}^*(M)$ and set $\operatorname{div}(m) = \sum_{i=1}^{n} n_i . z_i$. For $i = 1,\dots,n$ choose charts (U_i,ϕ_i) containing z_i and such that $\phi_i(U_i)$ contains $\bar{D}_1(0)$. Set $D_i = \phi_i^{-1}(\bar{D}_1(0))$ and $\gamma_i = \partial D_i = \phi_i^{-1}(\partial\bar{D}_1(0))$. We shall assume that the charts are chosen so that the sets D_i are mutually disjoint.

Log m is defined up to integer multiples of $2\pi i$ on $M' = M \setminus \bigcup_i D_i$. Hence the 1-form $\phi = d(\log m)$ is well defined on M'. By Stokes' theorem

$$\int_{M'} d\phi = \sum_{i=1}^{n} \int_{\gamma_i} \phi .$$

But $d\phi = d^2(\log m) = 0$, and so the left hand integral must vanish. Now

$$\sum_{i=1}^{n} \int_{\gamma_i} \phi = \sum_{i=1}^{n} \int_{|z|=1} \partial(\log m_i)/\partial z \, dz, \text{ where } m_i = m\phi_i^{-1}$$

$$= \sum_{i=1}^{n} \int_{|z|=1} m_i'/m_i \, dz$$

$$= \sum_{i=1}^{n} n_i, \text{ by the residue theorem.}$$

Hence $\deg(d) = 0$. □

The reader is warned that the vanishing of the degree of d is *not* generally sufficient for d to be the divisor of a meromorphic function (see the end of §5 and also Chapter 4, §4).

In view of the importance of Theorem 1.3.1 in the proofs of the Mittag-Leffler and Weierstrass theorems it is natural to seek a generalisation to Riemann surfaces. The first difficulty we encounter is that of finding a suitable definition of $\partial/\partial\bar{z}$ for a Riemann surface. Indeed, suppose $f \in C^\infty(M)$ and (U,ϕ) and (V,ψ) are complex analytic charts on M.

In general, $[\partial(f\phi^{-1})/\partial\bar{z}]\phi \neq [\partial(f\psi^{-1})/\partial\bar{z}]\psi$ and so we cannot expect to define $\partial/\partial\bar{z}$ as an endomorphism of $C^{\infty}(M)$. The correct generalisation of $\partial/\partial\bar{z}$ involves the introduction of "twisted" functions and tensor fields on M. These concepts are most easily discussed in the general framework of vector bundles and we shall defer further consideration of the construction of meromorphic functions until we have developed sufficient of the elementary theory of vector bundles in §5. The reader who is largely unfamiliar with the theory of vector bundles may prefer to go straight to Chapter 2 and return to section 5 after the end of Chapter 4.

Exercises.

1. Construct the (compact) Riemann surface of $z^2 = (y-a)(y-b)$, $a \neq b$, and show that it is biholomorphic to the Riemann sphere. (Hint: prove that the Riemann surface is simply connected).

2. Construct the Riemann surface of $y^2 = (z-a)(z-b)(z-c)$, a,b,c distinct, and show that it is diffeomorphic to the two dimensional real torus.

3. Formalise and prove a version of the Mittag-Leffler theorem valid for the Riemann sphere.

4. Let $f: M \to N$ be a holomorphic map between Riemann surfaces. Show

   a) Either f is constant or $f(M)$ is an open subset of N.

   b) In case M is compact, either f is constant or $f(M) = N$.

   c) Given $z_0 \in N$, $f^{-1}(z_0)$ is a discrete subset of M (possibly empty).

5. Let $f: M \to N$ be a non-constant map between Riemann surfaces M and N and suppose $d \in \mathcal{D}(N)$. Show how to define $f^*(d) \in \mathcal{D}(M)$ and verify that if $m \in \mathcal{M}^*(N)$ then $\mathrm{div}(mf) = f^*(\mathrm{div}(m))$.

§5. Vector bundles

The first part of this section constitutes a summary of the theory of real and complex vector bundles over a topological space. For more comprehensive treatments the reader may refer to Abraham and Marsden [1], Husmoller [1] or Lang [1].

Definition 1.5.1. Let X be a topological space. An *n-dimensional real vector bundle* E over X consists of a topological space E and continuous map $\pi: E \to X$ which satisfy

a) The fibre $\pi^{-1}(x) = E_x$ has the structure of a real n-dimensional vector space for every $x \in X$.

b) There exists an open cover $\{U_i: i \in I\}$ of X and homeomorphisms $\theta_i: \pi^{-1}(U_i) \to U_i \times \mathbb{R}^n$ such that for all $i \in I$, $x \in U_i$, θ_i maps E_x linearly and isomorphically onto $\{x\} \times \mathbb{R}^n$.

We denote the vector bundle by $\pi: E \to X$ or just E.

The topological space E is called the *total* space of the vector bundle, X is called the *base space* and π the *projection map*. The maps θ_i are called *trivialisations* of E.

Example 1. Let $\pi: X \times \mathbb{R}^n \to X$ denote projection on the first factor. Then $\pi: X \times \mathbb{R}^n \to X$ is an n-dimensional vector bundle over X called the *trivial* n-dimensional vector bundle over X. We often denote the trivial bundle over X with fibre $\mathbb{R}^n$ by $\underline{\mathbb{R}}^n$.

Condition b) on a vector bundle implies that a vector bundle is locally a trivial bundle. Globally it may be twisted.

Definition 1.5.2. A (continuous) *section* of the vector bundle $\pi: E \to X$ is a (continuous) map $s: X \to E$ such that πs = identity. That is, $s(x) \in E_x$ for all $x \in X$.

Example. If $f: X \to \mathbb{R}^n$, then the graph map $x: \to (x, f(x))$ is a section of the trivial bundle $X \times \mathbb{R}^n$. Conversely, every section of the trivial bundle $X \times \mathbb{R}^n$ defines an $\mathbb{R}^n$-valued function on X.

We shall let $C^0(E)$ denote the set of all continuous sections of E and remark that $C^0(E)$ has the structure of a real vector space with addition and scalar multiplication induced from the vector space structure on the fibres of E.

If s is a section of the vector bundle E, then $\theta_i s: U_i \to U_i \times \mathbb{R}^n$ is the graph map of a function $S_i: U_i \to \mathbb{R}^n$. Thus we may think of sections of E as *locally* being $\mathbb{R}^n$-valued functions. Globally, they may be "twisted" $\mathbb{R}^n$-valued functions.

Example. Let $S^1 \dot{\times} \mathbb{R}$ denote the twisted cylinder. $S^1 \dot{\times} \mathbb{R}$ is homeomorphic to the Mobius band and may be explicitly parametrized as the subset $\{(\cos\theta, \sin\theta, t\cos\theta/2, t\sin\theta/2): \theta \in [0,2\pi), t \in \mathbb{R}\}$ of $\mathbb{R}^4$. The space $S^1 \dot{\times} \mathbb{R}$ has the structure of a 1-dimensional real vector bundle over S^1 if we define projection onto S^1 in the obvious way.

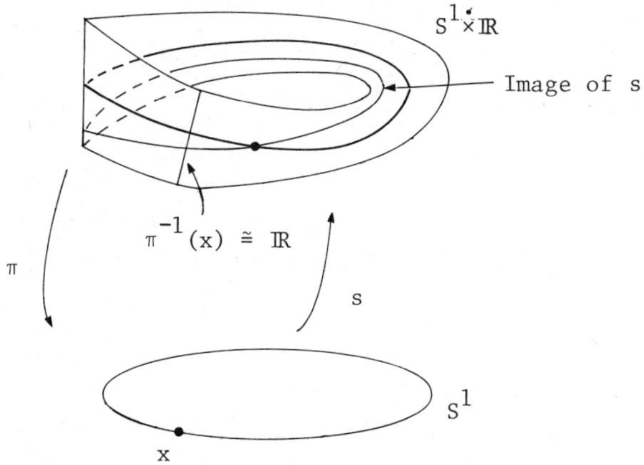

Figure 1.

We observe that every continuous section of $S^1 \dot{\times} \mathbb{R}$ is zero at at least one point in sharp contrast to what happens for sections of $S^1 \times \mathbb{R}$. Thus although *locally*, twisted and untwisted functions on S^1 possess the same properties, their global behaviour may be rather different.

We shall now give an alternative description of vector bundles in terms of *transition functions*. We follow the notation of Definition 1.4.4. for each $i \in I$, $x \in U_i$, we let

$$\theta_{ix}: E_x \to \mathbb{R}^n$$

denote the restriction of θ_i to E_x composed with the projection on $\mathbb{R}^n$.

Define

$$\theta_{ij}: U_{ij} \to GL(\mathbb{R}^n)$$

by $\theta_{ij}(x) = \theta_{ix}\theta_{jx}^{-1}$, $x \in U_{ij}$, $i,j \in I$ ($GL(\mathbb{R}^n)$ denotes the group of linear isomorphisms of $\mathbb{R}^n$ which is commonly given the alternative notation $GL(n,\mathbb{R})$.

It may easily be verified that the θ_{ij} are continuous maps which satisfy the cocycle conditions

$$\left.\begin{aligned} \theta_{ij} &= \theta_{ji}^{-1} \\ \theta_{ij}.\theta_{jk} &= \theta_{ik} \end{aligned}\right\} \qquad \ldots.A$$

(Inversion and multiplication are in $GL(\mathbb{R}^n)$). The θ_{ij} are called *transition functions* of the vector bundle E. Suppose s is a section of E and $S_i\colon U_i \to \mathbb{R}^n$ is the local representation of s on U_i as described above. Then for all i,j

$$\theta_{ij}S_j = S_i \text{ on } U_{ij} \qquad \ldots.B$$

Conversely, any set of functions $S_i\colon U_i \to \mathbb{R}^n$ satisfying B defines a unique section of E.

If E and F are vector bundles over X we shall say that a continuous map $A\colon E \to F$ is a *vector bundle map* if for all $x \in X$, $A(E_x) \subseteq F_x$ and $A|E_x$ is linear. If A is bijective and A^{-1} is a vector bundle map we shall say that A is a *vector bundle isomorphism* and that E and F are *isomorphic vector bundles.*

Suppose that E, F have trivialising maps θ_i, η_i respectively, relative to some (common) cover $\{U_i\colon i \in I\}$ of X. Then for each $i \in I$, A induces a continuous map $A_i\colon U_i \times \mathbb{R}^n \to U_i \times \mathbb{R}^n$ given by $A_i = \eta_i A \theta_i^{-1}$. In turn, A_i induces a continuous map

$$a_i\colon U_i \to L(\mathbb{R}^n,\mathbb{R}^n)$$

such that for all i,j

$$\eta_{ij}a_j = a_i\theta_{ij} \text{ on } U_{ij} .$$

Conversely, a family of continuous maps $a_i: U_i \to L(\mathbb{R}^n, \mathbb{R}^m)$ determines a vector bundle map from E to F provided that for all i,j

$$\eta_{ij} a_j = a_i \theta_{ij} \text{ on } U_{ij} . \qquad \ldots\ldots C$$

Notice that E will be a trivial vector bundle (that is, isomorphic to a trivial vector bundle) if and only if there exist $a_i: U_i \to GL(\mathbb{R}^n)$ such that for all i,j

$$a_j = a_i \theta_{ij} \text{ on } U_{ij} .$$

Suppose that $\{U_i: i \in I\}$ is an open cover of the topological space X and that we are given continuous maps $\theta_{ij}: U_{ij} \to GL(\mathbb{R}^n)$ satisfying the cocycle condition A. Then the θ_{ij} determine a unique - up to isomorphism - n-dimensional real vector bundle over X whose transition functions are the θ_{ij}. To see this we let Z denote the disjoint union over I of the products $U_i \times \mathbb{R}^n$. We define an equivalence relation on Z by $(i,x,U) \sim (j,y,V)$ iff $x = y$ and $U = \theta_{ij}(x)V$. It is an easy matter to show that $Z/\sim$ has the structure of an n-dimensional real vector bundle over X with transition functions θ_{ij} and we leave details to the reader. In the sequel we usually construct vector bundles by specifying a set of transition functions. From C above, it follows that two sets $\theta_{ij}, \eta_{ij}: U_{ij} \to GL(\mathbb{R}^n)$ of transition functions define isomorphic vector bundles if and only if there exist $a_i: U_i \to GL(\mathbb{R}^n)$ such that for all i,j, $\eta_{ij} a_j = a_i \theta_{ij}$ on U_{ij}.

Notation. Let $\mathbb{R}^{m'}$ denote the real dual of $\mathbb{R}^m$. If $A \in L(\mathbb{R}^m, \mathbb{R}^n)$, then the *transpose* of A, A', is the linear map from $\mathbb{R}^{n'} \to \mathbb{R}^{m'}$ characterised by $A'(\phi)(e) = \phi(A(e))$, $\phi \in \mathbb{R}^{n'}$, $e \in \mathbb{R}^m$.

Example 4. Let E be an n-dimensional real vector bundle with transition functions θ_{ij}. We define $\theta'_{ij}: U_{ij} \to GL(\mathbb{R}^{n'})$ by

$$\theta'_{ij}(x) = [\theta_{ij}(x)']^{-1}, \ x \in U_{ij}.$$

The θ'_{ij} satisfy the cocycle condition A and so are the transition functions for a vector bundle which we call the *dual* vector bundle of E and denote by E'. We remark that the trivialisations θ'_i of E' map naturally to $U_i \times \mathbb{R}^{n'}$ (rather than $U_i \times \mathbb{R}^n$).

Motivated by the previous example, we shall now extend the possible "fibre models" of real vector bundles to include any finite combination of direct sum, tensor product, exterior and symmetric power of $\mathbb{R}^n$ and $\mathbb{R}^{n'}$. All these vector space operations extend immediately to vector bundles over a fixed base space X. We give one example to illustrate the method and refer the reader to the references for more details. Suppose that E and F are vector bundles over X with transition functions $\theta_{ij}: U_{ij} \to GL(\mathbb{R}^m)$ and $\eta_{ij}: U_{ij} \to GL(\mathbb{R}^n)$ respectively. $\wedge^r F \otimes (E \oplus (\odot^s F'))$ will then denote the vector bundle over X with fibre model $\wedge^r \mathbb{R}^n \otimes (\mathbb{R}^m \oplus (\odot^r \mathbb{R}^{n'}))$ and transition functions $\psi_{ij}: U_{ij} \to GL(\wedge^r \mathbb{R}^n \otimes (\mathbb{R}^n \oplus (\odot^r \mathbb{R}^{n'})))$ defined by

$$\psi_{ij}(x) = (\wedge^r \eta_{ij}(x)) \otimes (\theta_{ij}(x) \oplus (\odot^s \eta'_{ij}(x))), \ x \in U_{ij}.$$

($\wedge^r$ and $\odot^s$ denote the operations of pth. exterior and sth. symmetric power respectively.

Suppose that X is a differential manifold (with C^∞ structure). We can define smooth (that is C^∞) real vector bundles over X in the obvious way and everything we have said above generalises immediately to the smooth case. If E is a smooth vector bundle over X, we let $C^r(E)$ denote the vector space of C^r sections of E, $1 \le r \le \infty$.

Example 5. Let M be an m-dimensional differential manifold with smooth atlas $\{(U_i, \phi_i): i \in I\}$. We define C^∞ maps $\phi_{ij}: U_{ij} \to GL(\mathbb{R}^m)$ by $\phi_{ij}(x) = D(\phi_i \phi_j^{-1})(\phi_j(x))$. The ϕ_{ij} are the transition functions for the (real) *tangent bundle* of M which we denote in the sequel by $\mathcal{T}M$. The (real) *cotangent bundle* of M, $\mathcal{T}'M$, is the dual bundle of $\mathcal{T}M$ with transition functions ϕ'_{ij} (See Abraham and Marsden [1], Chillingworth [1] and Lang [1] for more details).

We now turn our attention to vector bundles with fibre a complex vector space. An m-*dimensional complex vector bundle* E over X is defined exactly as in Definition 1.5.1 except that "real" and "$\mathbb{R}^n$" are everywhere replaced by "complex" and "$\mathbb{C}^m$" respectively. We leave the writing out of the formal definitions of complex vector bundles, sections, transition functions, etc. to the reader. Sections of an m-dimensional complex vector bundle will locally be $\mathbb{C}^m$-valued functions: globally they may be twisted.

One important feature of complex vector bundles is the greater complexity of the fibre models which involve not only duals but also conjugates. But first we need to review some complex linear algebra.

If $J: \mathbb{C}^m \to \mathbb{C}^m$ denotes scalar multiplication by i, then J is a $\mathbb{C}$-linear isomorphism of $\mathbb{C}^m$ satisfying $J^2 = -I$. The usual complex structure on $\mathbb{C}^m$ may be defined in terms of J and the underlying real structure on $\mathbb{C}^m$ by

$$(a+ib)Z = aZ + bJ(Z), \quad a,b \in \mathbb{R}, \quad Z \in \mathbb{C}^m .$$

The *conjugate complex structure* on $\mathbb{C}^m$ is defined by

$$(a+ib)Z = aZ - bJ(Z), \quad a,b \in \mathbb{R}, \quad Z \in \mathbb{C}^m .$$

We denote the resulting complex vector space by $\bar{\mathbb{C}}^m$. Suppose $A \in L(\mathbb{C}^m,\mathbb{C}^n)$, we define $\bar{A} \in L(\bar{\mathbb{C}}^m,\bar{\mathbb{C}}^n)$ by simply requiring that $A = \bar{A}$ on the common underlying set of $\mathbb{C}^m$ and $\bar{\mathbb{C}}^m$ (We should emphasise that all linear maps here are assumed *complex* linear unless the contrary is explicitly stated).

We let $\mathbb{C}^{m*} = L(\mathbb{C}^m,\mathbb{C})$ denote the *conjugate dual space* of $\mathbb{C}^m$. If $A \in L(\mathbb{C}^m,\mathbb{C}^n)$, we define the *transpose* $A^* \in L(\mathbb{C}^{n*},\mathbb{C}^{m*})$ exactly as in the real case.

We let $\bar{\mathbb{C}}^{m*}$ denote the complex vector space of conjugate complex linear maps on $\mathbb{C}^m$. We call $\bar{\mathbb{C}}^{m*}$ the *conjugate dual space* of $\mathbb{C}^m$. Note that $f \in \bar{\mathbb{C}}^{m*}$ if $f: \mathbb{C}^m \to \mathbb{C}$ is $\mathbb{R}$-linear and $f(\lambda Z) = \bar{\lambda} f(Z)$, $\lambda \in \mathbb{C}$, $Z \in \mathbb{C}^m$. Clearly, $\bar{\mathbb{C}}^{m*} \approx L(\bar{\mathbb{C}}^m,\mathbb{C})$. Moreover, $\overline{\mathbb{C}^{m*}} \approx \bar{\mathbb{C}}^{m*}$ where we map $\phi \in \mathbb{C}^{m*}$ to $\bar{\phi} \in \bar{\mathbb{C}}^{m*}$ and $\bar{\phi}(e) = \overline{\phi(e)}$, $e \in \mathbb{C}^m$. If $A \in L(\mathbb{C}^m,\mathbb{C}^n)$, we define the *conjugate transpose* $\bar{A}^* \in L(\bar{\mathbb{C}}^{n*},\bar{\mathbb{C}}^{m*})$ by

$$\bar{A}^*(\phi)(e) = \phi(A(e)), \quad e \in \mathbb{C}^m, \quad \phi \in \bar{\mathbb{C}}^{n*} .$$

Finally, if $A \in L(\mathbb{C}^m,\mathbb{C}^n)$ has matrix $[a_{ij}]$ relative to the standard bases of $\mathbb{C}^m$ and $\mathbb{C}^n$, the reader may easily verify that the matrices of $\bar{A}$, A^* and $\bar{A}^*$ are respectively $[\overline{a_{ij}}]$, $[a_{ji}]$ and $[\overline{a_{ji}}]$ relative to the naturally induced bases on $\bar{\mathbb{C}}^m,\ldots,\bar{\mathbb{C}}^{n*}$.

Example 6. Let E be a complex vector bundle with transition functions $\theta_{ij}: U_{ij} \to GL(\mathbb{C}^m)$. We define maps

$$\bar{\theta}_{ij}: U_{ij} \to GL(\bar{\mathbb{C}}^m)$$

$$\theta^*_{ij}: U_{ij} \to GL(\mathbb{C}^{m*}),$$

$$\bar{\theta}^*_{ij}: U_{ij} \to GL(\bar{\mathbb{C}}^{m*})$$

by $\bar{\theta}_{ij}(x) = \overline{\theta_{ij}(x)}$, $\theta^*_{ij}(x) = [\theta_{ij}(x)^*]^{-1}$, $\bar{\theta}^*_{ij}(x) = [\overline{\theta_{ij}(x)^*}]^{-1}$, $x \in U_{ij}$.

Now $\bar{\theta}_{ij}$, θ^*_{ij}, $\bar{\theta}^*_{ij}$ satisfy the cocycle condition A and so are the transition functions for complex vector bundles which we call the *conjugate* bundle, $\bar{E}$, *dual* bundle, E^*, and *conjugate dual* bundle, $\bar{E}^*$, of E respectively.

Exactly as for the real case we may now extend the possible fibre models of complex vector bundles to include finite combinations of direct sum, tensor product, exterior and symmetric powers of $\mathbb{C}^m$ and its conjugate and dual spaces. All these operations extend immediately to complex vector bundles over a fixed base space and we leave formal details to the reader.

Remarks.

1. We shall show in Chapter 9 that, as complex vector bundles, E is isomorphic to $\bar{E}^*$ and E^* is isomorphic to $\bar{E}$. However, these isomorphisms are *not* natural and in the sequel it will often be convenient to regard these bundles as all being distinct.

2. If E is a 1-dimensional complex vector bundle, we shall usually refer to E as a *complex line bundle*. Since $GL(\mathbb{C}) \approx \mathbb{C}^{\bullet}$, the transition functions for a complex line bundle may be equivalently given by non-vanishing complex valued functions: $\theta_{ij}: U_{ij} \to \mathbb{C}^{\bullet}$.

Examples.

7. Let CLB(X) denote the set of isomorphism classes of complex line bundles over X. We claim that CLB(X) has the natural structure of an abelian group with identity the trivial bundle $\underline{\mathbb{C}}$ and product and inversion respectively defined by

$$E.F = E \otimes F, \; E,F \in CLB(X)$$

$$E^{-1} = E^*, \; E \in CLB(X).$$

Certainly $E.\underline{\mathbb{C}} = \underline{\mathbb{C}}.E = E$ for all $E \in CLB(X)$. We verify that E^* is the inverse of E with respect to $\otimes$ and leave the remaining details to the reader. To verify that E^* is the inverse of E it is enough to note that there is a natural isomorphism $E \otimes E^* \approx \underline{\mathbb{C}}$ induced from the dual pairing $E_x \otimes E_x^* \approx \mathbb{C}$, $x \in X$ (Perhaps we should remark that for the group RLB(X) of *real* line bundles on X, L is isomorphic to L', $L \in RLB(X)$, and so $L^2 \approx \underline{\mathbb{R}}$, for all $L \in RLB(X)$).

2. Let M be a Riemann surface with complex analytic atlas $\{(U_i,\phi_i): i \in I\}$. We define $\phi_{ij}: U_{ij} \to GL(\mathbb{C})$ by $\phi_{ij}(z) = \partial(\phi_i\phi_j^{-1})/\partial z(\phi_j(z))$, $z \in U_{ij}$. The ϕ_{ij} are complex analytic maps and are the transition functions for the *holomorphic tangent bundle* of M which we denote in the sequel by TM.

$\phi_{ij}^*: U_{ij} \to GL(\mathbb{C}^*)$ are also analytic maps and define the transition functions for the *holomorphic cotangent bundle*, TM*, of M.

$\phi_{ij}: U_{ij} \to GL(\bar{\mathbb{C}})$ are C^∞ maps (not analytic!) and are the transition functions for the *anti-holomorphic tangent bundle*, $\overline{TM}$, of M.

$\bar{\phi}_{ij}^*: U_{ij} \to GL(\bar{\mathbb{C}}^*)$ are C^∞ maps and are the transition functions for the *anti-holomorphic cotangent bundle*, $\overline{TM}^*$, of M.

We shall now show how the operator $\partial/\partial\bar{z}$ extends to a Riemann surface M as a map $\bar{\partial}: C^\infty(M) \to C^\infty(\overline{TM}^*)$. First we need to note the composite mapping formula for $\partial/\partial z$, $\partial/\partial\bar{z}$:

Suppose h and g are C^1 complex valued functions defined on open subsets of $\mathbb{C}$. Then

$$\partial(gh)/\partial z = \partial g/\partial z \; \partial h/\partial z + \partial g/\partial\bar{z} \; \overline{\partial h/\partial\bar{z}} \qquad \ldots *$$

$$\partial(gh)/\partial\bar{z} = \partial g/\partial z \; \partial h/\partial\bar{z} + \partial g/\partial\bar{z} \; \overline{\partial h/\partial z} \qquad \ldots **$$

The proof is most easily done by taking real and imaginary parts and applying the chain rule for $\partial/\partial x$ and $\partial/\partial y$.

Let (U_i,ϕ_i), (U_j,ϕ_j) be charts for M and $f \in C^\infty(M)$. Then

$$\begin{aligned}\partial(f\phi_i^{-1})/\partial\bar{z} &= \partial(f\phi_j^{-1}\phi_j\phi_i^{-1})/\partial\bar{z}\\ &= \partial(f\phi_j^{-1})/\partial\bar{z}\ \overline{\partial(\phi_j\phi_i^{-1})/\partial z}, \text{ using ** above}\\ &= \bar{\phi}^*_{ij}\ \partial(f\phi_j^{-1})/\partial\bar{z},\end{aligned}$$

where $\bar{\phi}^*_{ij}$ are the transition functions for $\overline{TM}^*$. Hence $\partial(f\phi_i^{-1})/\partial\bar{z}$ are the local representatives of a C^∞ section of $\overline{TM}^*$ which in the sequel we shall denote by $\bar{\partial}f$. The operator $\bar{\partial}$: $C^\infty(M) \to C^\infty(\overline{TM}^*)$ clearly has kernel equal to A(M). We shall return to the question of how far $\bar{\partial}$ fails to be surjective at the end of this section.

The fibre model for $\overline{TM}^*$ is $\bar{\mathbb{C}}^*$ and we denote the standard basis of $\bar{\mathbb{C}}^*$ by $d\bar{z}$. Suppose that $g \in C^\infty(\overline{TM}^*)$, then the local representative of g with respect to a chart (U_i,ϕ_i) is given by $g_i d\bar{z}$, where $g_i = \bar{\phi}_i^* g \in C^\infty(U_i)$. If $f \in C^\infty(M)$, the local representative of $\bar{\partial}f$ is $\partial f_i/\partial\bar{z}\ d\bar{z}$, where $f_i = f\phi_i^{-1} \in C^\infty(U_i)$.

We may similarly define ∂: $C^\infty(M) \to C^\infty(TM)$ and the local representative of ∂f with respect to a chart (U_i,ϕ_i) will be $\partial f_i/\partial z\ dz$, where dz denotes the standard basis of $\mathbb{C}^*$.

We remark that $df = \partial f + \bar{\partial} f$ as is easily verified by the local computation $\partial f/\partial z\ dz + \partial f/\partial\bar{z}\ d\bar{z} = \partial f/\partial x\ dx + \partial f/\partial y\ dy$.

Definition 1.5.3. Let θ_{ij}: $U_{ij} \to GL(\mathbb{C})$ be the transition functions for a complex line bundle E over the Riemann surface M. If the θ_{ij} are all analytic, we shall say that E is a *holomorphic line bundle* over M.

Examples.

9. If M is a Riemann surface, TM and TM* are holomorphic line bundles over M.

10. Let HLB(M) denote the set of isomorphism classes of holomorphic line bundles over the Riemann surface M (isomorphism here is, of course, analytic line bundle isomorphism, and may be given explicitly in terms of analytic functions using condition C above).

As in the previous example on CLB(M), HLB(M) has the natural structure of an abelian group with composition defined as tensor product and inverse as dual (see also below).

Holomorphic line bundles and their generalisation, holomorphic vector bundles, will be of the greatest importance in the sequel. As we shall see they provide a particularly potent geometric framework for many problems in complex analysis.

Suppose that E is a holomorphic line bundle on the Riemann surface M with transition functions θ_{ij} relative to the cover $\{U_i\}$. A family $S_i \in A(U_i)$ will define an *analytic section* of E if, for all i,j,

$$S_i = \theta_{ij} S_j \text{ on } U_{ij}.$$

A family $M_i \in \mathcal{M}(U_i)$ will define a *meromorphic section* m of E if, for all i,j,

$$M_i = \theta_{ij} M_j \text{ on } U_{ij} .$$

We denote the sets of analytic and meromorphic sections of E by $\Omega(E)$ and $\mathcal{M}(E)$ respectively. We let $\mathcal{M}^*(E)$ denote the set of non-zero meromorphic sections of E. As in §§3,4, we define ord(m,z), $m \in \mathcal{M}(E)$, $z \in M$; div(m), $m \in \mathcal{M}^*(E)$ and, if M is compact, deg(div(m)), $m \in \mathcal{M}^*(E)$.

With the above formalism out of the way we can now rapidly achieve one of the primary goals of this section: The geometric formulation of the problem of generalising Weierstrass' theroem to compact Riemann surfaces.

For the remainder of this section we shall assume that M is a compact Riemann surface.

Proposition 1.5.4. There is a natural homomorphism $[\ ]: \mathcal{D}(M) \to HLB(M)$ satisfying the property that for any $d \in \mathcal{D}(M)$ there exists $s(d) \in \mathcal{M}^*([d])$ such that div(s(d)) = d. Furthermore, s(d) is determined uniquely up to multiplication by elements of $\mathbb{C}^{\bullet}$.

Proof. As we showed in §3, any divisor may be uniquely specified by an open cover $\{U_i\}$ of M and $m_i \in \mathcal{M}^*(U_i)$ such that, for all i,j, $m_i/m_j \in A^*(U_{ij})$. Define $\theta_{ij} = m_i/m_j \in A^*(U_{ij})$. Clearly the θ_{ij} are the transition functions for a holomorphic line bundle on M which we denote by [d]. We leave it to the reader to verify that $d \to [d]$ is a group homomorphism.

We define $s(d)_i = m_i \in M^*(U_i)$ and observe that $\theta_{ij} s(d)_j = s(d)_i$ on U_{ij}. Hence the $s(d)_i$ are the local representatives of a meromorphic section s(d) of [d]. Obviously div(s(d)) = d. We leave the remaining assertions as an exercise - see also Lemma 1.5.6. □

Remark. Proposition 1.5.4 gives a "formal" solution to the problem of which divisors are divisors of meromorphic functions: Every divisor is the divisor of a "twisted" meromorphic function.

Proposition 1.5.5. A divisor d on M is the divisor of a meromorphic function if and only if [d] is trivial as a holomorphic line bundle.

Proof. Suppose $d = \text{div}(m)$, $m \in M^*(M)$. Let $\{U_i\}$ be any open cover of M. Set $m_i = m|U_i$ and $\theta_{ij} = m_i/m_j \equiv 1$. The m_i then define the divisor d and, since $\theta_{ij} \equiv 1$, [d] is holomorphically trivial. Conversely, if [d] is holomorphically trivial, there exists a nowhere zero analytic section s of [d]. In terms of local representatives $s_i \in A^*(U_i)$ we must have $\theta_{ij} s_i = s_j$, where θ_{ij} are the transition functions of [d]. But $\theta_{ij} = m_i/m_j$, where the m_i define the divisor d. Therefore, we have $m_i/s_i = m_j/s_j$ on U_{ij} and we may define $m \in M^*(M)$ by $m|U_i = m_i/s_i$. Obviously, div(m) = d. □

Lemma 1.5.6. Suppose that $L \in \text{HLB}(M)$ and $s,t \in M^*(L)$. Then s/t naturally defines a non-zero meromorphic function on M.

Proof. Suppose that L has transition functions θ_{ij} relative to some open cover $\{U_i\}$ of M. Then, s, t are locally given by $s_i, t_i \in M^*(U_i)$ subject to the relations $\theta_{ij} s_{ij} = s_i$, $\theta_{ij} t_j = t_i$ for all i,j. Therefore, $s_j/t_j = s_i/t_i$ on U_{ij} and so we may define $m \in M^*(M)$ by $m|U_i = s_i/t_i$. □

Remark. Notice that t^{-1} naturally defines a meromorphic section of $L^{-1} = L^*$ and so s/t may equivalently be defined by using the pairing $M^*(L) \times M^*(L^{-1}) \to M^*(M)$ induced from the dual pairing $L \otimes L^* \approx \underline{\mathbb{C}}$.

Proposition 1.5.7. Let $L \in HLB(M)$ and suppose $\mathcal{M}^*(L) \neq \emptyset$. Given $s \in \mathcal{M}^*(L)$, set $\deg(s) = \deg(\text{div}(s))$. Then $\deg(s)$ is independent of s and depends only on L.

Proof. Let $s,t \in \mathcal{M}^*(L)$. Clearly, $\text{div}(s/t) = \text{div}(s) - \text{div}(t)$. But $s/t \in \mathcal{M}^*(M)$ and so by Lemma 1.4.3, $\deg(s/t) = 0$. Hence $\deg(s) = \deg(t)$. □

For every holomorphic line bundle L on M that admits a non-trivial meromorphic section we may define the *degree of L*, deg(L), to be deg(s), where s is any non-trivial meromorphic section of L. Notice that if $\deg(L) \neq 0$, then L cannot be holomorphically trivial. It is also clear that if $d \in \mathcal{D}(M)$ then $\deg(d) = \deg([d])$.

The degree of L is an important topological invariant of L. In fact we shall show later (Example 14, §3, Chapter 6) that $\deg(L) = 0$ if and only if L is trivial as a *complex* line bundle.

What we have achieved above is to shift the problem of determining which divisors are divisors of meromorphic functions to the apparently more general problem of determining which holomorphic line bundles admit non-trivial meromorphic sections. We already know that a divisor d can be the divisor of a meromorphic function on M only if $\deg([d]) = \deg(d) = 0$. Let $HLB_0(M)$ denote the subgroup of HLB(M) consisting of line bundles of degree zero. Two problems naturally arise at this stage.

1. Show that $[\ ]: \mathcal{D}(M) \to HLB(M)$ is onto.
2. Describe $HLB_0(M)$.

To show that [ ] is onto amounts to proving that every holomorphic line bundle on M has a non-trivial meromorphic section. The existence of such meromorphic sections follows from a rather general finiteness theorem that we prove in Chapter 7 (see the exercises at the end of §5, Chapter 7). Once we know that [ ] is onto it is an easy matter to prove Riemann's part of the *Riemann-Roch* theorem:

For any divisor d on M we have

$$\dim_{\mathbb{C}} \Omega([d]) \geq \deg(d) + 1 - g,$$

where g denotes the genus of M.

This result is already sufficient to prove useful existence theorems for meromorphic functions on compact Riemann surfaces. The full Riemann-Roch theorem depends on Serre-duality which we establish in Chapter 10 (see also § 5, Chapter 7).

As far as the second problem goes, $HLB_0(M)$ may be shown to have the structure of a compact complex Lie group of complex dimension g. The group $HLB_0(M)$ is usually referred to as the *Picard variety* of M and denoted by Pic(M). Whilst the dimension of Pic(M) depends only on the genus on M its complex structure depends on the complex structure of M. Abel's theorem gives explicit conditions on a divisor in order that it determine the identity of Pic(M), that is the trivial holomorphic line bundle (We refer to Gunning [1] for details).

We conclude this section by briefly indicating the role that $\bar{\partial}$ plays in the study of meromorphic functions on a compact Riemann surface.

Suppose that E is a holomorphic line bundle on M with transition functions θ_{ij}. We shall investigate conditions under which E is holomorphically trivial. First observe that, as in the proof of Theorem 1.3.3, $\{(2\pi i)^{-1}\log\theta_{ij}\}$ defines a class γ_E in $H^2(M, \mathbb{Z})$. The class γ_E will be zero if and only if E is trivial as a complex line bundle. From now on we assume E is complex trivial, that is $E \in \mathrm{Pic}(M)$. We may choose the branches of $\log \theta_{ij}$ so that $\log \theta_{ij} = h_{ij} \in A(U_{ij})$ determine an additive cocycle (see the proof of Theorem 1.3.3). Now E is trivial as a complex line bundle if there exist $g_i \in C^\infty(U_i)$ such that $h_{ij} = g_j - g_i$ (exponentiate!). Deifne $F \in C^\infty(\overline{TM^*})$ by $F|U_i = \bar{\partial} g_i$. We can solve $\bar{\partial} u = -F$ if the class of F in $C^\infty(\overline{TM^*})/\bar{\partial} C^\infty(M)$ is zero. Given such a solution, the functions $a_i = \exp(g_i + u) \in A^*(U_i)$ will then give a holomorphic trivialisation of E. The basic result now is that the complex vector space $C^\infty(\overline{TM^*})/\bar{\partial} C^\infty(M)$ is finite dimensional (the dimension is actually g, the genus of M). If we are more careful in our arguments we see that to prove E is holomorphically trivial it is enough to find $u \in C^\infty(M)$ such that $\bar{\partial} u + F$ exponentiates to zero. Further analysis then shows that $H^1(M,\mathbb{Z})$ ($\cong \mathbb{Z}^{2g}$) determines an integer lattice in $C^\infty(\overline{TM^*})/\bar{\partial} C^\infty(M)$ and that a necessary and sufficient condition for E to be holomorphically trivial is that F determines a lattice point. The Picard variety is then isomorphic to the torus $(C^\infty(\overline{TM^*})/\bar{\partial} C^\infty(M))/H^1(M,\mathbb{Z})$. We refer to Gunning [1] for details. In Chapter 7 we shall show that $C^\infty(\overline{TM^*})/\bar{\partial}C^\infty(M)$ is finite dimensional using rather general arguments about sheaves. In Chapter 10 we give a more direct proof of finiteness and much more explicit description of the quotient $C^\infty(\overline{TM^*})/\bar{\partial}C^\infty(M)$ using Hodge theory.

Exercises.

1. Let $f: X \to Y$ be a continuous map between topological spaces and E be a vector bundle on Y with transition functions θ_{ij} relative to some cover $\{U_i\}$ of Y. Show that $\theta_{ij}f$ are transition functions for a vector bundle f^*E over X (the "pull-back bundle of E by f"). Verify that, up to vector bundle isomorphism, f^*E depends only on E and f and not on any choices we have made. Show also that in case f is a holomorphic map between Riemann surfaces our construction determines a homomorphism $f^*: HLB(Y) \to HLB(X)$.

2. Let $f: X \to Y$ be a holomorphic map between Riemann surfaces and let $d \in \mathcal{D}(Y)$. Prove that $f^*[d] \cong [f^*(d)]$ (see Exercise 5, §4).

3. Let M be a Riemann surface and $d, d' \in \mathcal{D}(M)$. We say d and d' are *linearly equivalent* if $d - d' = \mathrm{div}(m)$ for some $m \in M^*(M)$. Show that d and d' are linearly equivalent if and only if $[d] \cong [d']$ as holomorphic line bundles.

4. Let Ω be a domain in $\mathbb{C}$. Prove that every holomorphic line bundle on Ω is holomorphically trivial.

5. Let E, F be vector bundles over the differential manifold M. Prove that $C^\infty(E' \otimes F)$ is naturally isomorphic to the space of smooth vector bundle maps from E to F, Hom(E,F).

6. Suppose that E, F are vector bundles over the differential manifold M and $\phi: C^\infty(E) \to C^\infty(F)$ is a $C^\infty_{\mathbb{R}}(M)$-linear map. Prove that ϕ is induced from a vector bundle map $\Phi: E \to F$. That is, show that $\phi(s)(x) = \Phi(s(x))$, $x \in M$, $s \in C^\infty(E)$. Deduce that ϕ determines a unique section γ of $C^\infty(E' \otimes F)$ characterised by $\langle \gamma, X \rangle = \phi(X)$, $X \in C^\infty(E)$, where $\langle\ ,\ \rangle$ is induced from the dual pairing between E and E'.

Appendix to Chapter 1.

In the first part of this appendix we prove a number of results related to Cauchy's integral formula that will be used later in the text as well as in the proof of Theorem 1.3.1. The main element in the proof of Theorem 1.3.1 is the Runge approximation theorem and this, together with the proof of Theorem 1.3.1, constitute the remainder of the appendix.

In the statement of the next three results we shall assume that Ω is a bounded open domain in $\mathbb{C}$ with $\partial\Omega$ a finite union of piecewise C^1 curves. We use the notation $C^1(\bar{\Omega})$ to denote the set of functions on $\bar{\Omega}$ which are restrictions of C^1 functions defined on some neighbourhood of $\bar{\Omega}$.

Lemma A1.1. Let $f \in C^1(\bar{\Omega})$. Then

$$\int_{\partial\Omega} f dz = \int_{\Omega} \partial f/\partial\bar{z}\, d\bar{z}dz = 2i \int_{\Omega} \partial f/\partial\bar{z}\, dxdy .$$

Proof. Stoke's or Green's theorem in the plane (take real and imaginary parts). □

Theorem A1.2 (Cauchy's Integral Formula). Let $u \in C^1(\bar{\Omega})$. Then

$$u(\zeta) = (2\pi i)^{-1} \int_{\partial\Omega} \frac{u(z)}{z-\zeta} dz + \frac{1}{\pi} \int_{\Omega} \partial u/\partial\bar{z}/(z-\zeta) dxdy , \quad \zeta \in \Omega.$$

Proof. Let $\Omega_\varepsilon = \{z \in \Omega : |z-\zeta| > \varepsilon\}$, where $\varepsilon < d(\zeta, \partial\Omega)$. Apply Lemma A1.1 with $f(z) = u(z)/(z-\zeta)$ and let $\varepsilon \to 0$. □

Corollary A1.3. Let $u \in C^0(\bar{\Omega})$ and suppose u is analytic in Ω. Then

$$u(\zeta) = (2\pi i)^{-1} \int_{\partial\Omega} \frac{u(z)}{z-\zeta} dz, \quad \zeta \in \Omega.$$

Corollary A1.4. Let Ω be a domain in $\mathbb{C}$ and suppose that u_n, $n \geq 1$, is a sequence in $A(\Omega)$ which converges uniformly on compact subsets of Ω to a function u on Ω. Then $u \in A(\Omega)$.

Proof. Apply Corollary A1.3 to $u_n - u_m$ restricted to closed discs contained in Ω to deduce that $\partial u_n/\partial z$ converges uniformly on compact subsets of Ω. Since $\partial u_n/\partial\bar{z} = 0$, it follows that u is C^1 on Ω with $\partial u/\partial\bar{z} = 0$. □

Corollary A1.5 (Montel's theorem). Let Ω be a domain in $\mathbb{C}$ and suppose that u_n, $n \geq 1$, is a sequence in $A(\Omega)$ which is uniformly bounded on every compact subset of Ω. Then there is a subsequence of the u_n which converges uniformly on compact subsets of Ω to a limit $u \in A(\Omega)$.

Proof. By Cauchy's inequalities (Exercise 2, §1) the sequence $\partial u_n/\partial z$ is uniformly bounded on any closed disc $D_r(z) \subset \Omega$. Hence $\partial u_n/\partial z$ is uniformly bounded on any compact subset of Ω. Therefore the sequence u_n is an equicontinuous family and, by Ascoli's theorem, has a subsequence converging uniformly on compact subsets of Ω to a function u. By Corollary A1.4, $u \in A(\Omega)$. □

Let μ be a (finite, complex regular, Borel) measure on $\mathbb{C}$. The *support* of μ, $\mathrm{supp}(\mu)$, is defined to be the smallest closed subset of $\mathbb{C}$ outside of which μ is zero.

Theorem A1.6. Let μ be a measure on $\mathbb{C}$ with compact support. Then

$$u(\zeta) = \int_{\mathbb{C}} (z-\zeta)^{-1} d\mu(z)$$

defines an analytic function outside $\mathrm{supp}(\mu)$. If $d\mu = \frac{1}{\pi}\phi\, dxdy$, $\phi \in C_c^\infty(\mathbb{C})$, then $u \in C^\infty(\mathbb{C})$ and $\partial u/\partial \bar{z} = \phi$.

Proof. Certainly u is C^1 outside $\mathrm{supp}(\mu)$ - in fact C^∞ - and differentiating under the integral sign we see that $\partial u/\partial \bar{z} = 0$. Hence u is analytic outside $\mathrm{supp}(\mu)$. For the second statement we see by changing variables that

$$u(\zeta) = -\frac{1}{\pi}\int_{\mathbb{C}} \phi(\zeta - z) z^{-1} dxdy .$$

Now z^{-1} is integrable on compact sets and so $u \in C^\infty(\mathbb{C})$. Moreover,

$$\begin{aligned}\partial u/\partial\bar{\zeta} &= \frac{1}{\pi}\int_{\mathbb{C}} \partial\phi/\partial\bar{\zeta}(\zeta - z)/z \; dxdy \\ &= \frac{1}{\pi}\int_{\mathbb{C}} \partial\phi/\partial\bar{\zeta}(z)/(\zeta - z)\; dxdy .\end{aligned}$$

Apply Theorem A1.2 with Ω any disc containing $\mathrm{supp}(\mu)$ to deduce that $\partial u/\partial\bar{\zeta} = \phi$. □

Remark. Theorem A1.6 proves Theorem 1.3.1 in the special case that the right hand side is a C^∞ function with compact support. The solution u given by Theorem A1.6 need not have compact support as is easily seen by choosing ϕ so that $\int_{\mathbb{C}} \phi \neq 0$. This is in sharp distinction to what happens for more than one variable as we shall see in Chapter 2.

We shall now prove a special case of the Runge Approximation Theorem. The proof we give is modelled on that given in Gamelin [1] and Hörmander [1] and uses the Riesz representation theorem in combination with the Hahn-Banach theorem. An alternative proof may be found in Hille [1].

Theorem A1.7. Let Ω be an open subset of $\mathbb{C}$ and suppose that K is a compact subset of Ω such that $\Omega \setminus K$ has no component which is relatively compact in Ω. Then every function which is analytic on a neighbourhood of K can be uniformly approximated on K by functions in $A(\Omega)$.

Proof. By the Hahn-Banach and Riesz representation theorems, it is sufficient to show that every measure μ on K which is orthogonal to $A(\Omega)$ is also orthogonal to every function analytic on a neighbourhood of K.

Suppose f is analytic on a neighbourhood of K. Since we are interested in uniform approximation *on* K we can assume that $f \in C_c^\infty(\mathbb{C})$ (multiply f by $\phi \in C_c^\infty(\mathbb{C})$ where $\phi \equiv 1$ on K). By Theorem A1.2,

$$\begin{aligned} f(\zeta) &= \frac{1}{\pi}\int_{\mathbb{C}} \partial f/\partial\bar{z}\,(z-\zeta)^{-1}\,dxdy \\ &= \frac{1}{\pi}\int_{\mathbb{C}\setminus K} \partial f/\partial\bar{z}(z-\zeta)^{-1}\,dxdy\,, \end{aligned}$$

since f is analytic on a neighbourhood of K.

By Fubini's theorem

$$\int_{\mathbb{C}} f(\zeta)d\mu(\zeta) = \frac{1}{\pi}\int_{\mathbb{C}\setminus K} \partial f/\partial\bar{z}\left[\int_{\mathbb{C}} (z-\zeta)^{-1}d\mu(\zeta)\right]dxdy\,.$$

Set $u(z) = \int_{\mathbb{C}} (z-\zeta)^{-1}d\mu(\zeta)$. Then u is analytic on $\mathbb{C}\setminus K$, Theorem A1.6. We claim that $u \equiv 0$ on $\mathbb{C}\setminus K$. If so, it then follows that $\int_{\mathbb{C}} f\,d\mu = 0$ and the proof is complete.

Let $R = \sup\{|z| : z \in K\}$. Then $(z - \zeta)^{-1}$ can be expanded as a power series in ζ^n which converges uniformly on K provided that $|z| > R$. By our orthogonality assumption on μ, $\int_{\mathbb{C}} \zeta^n d\mu(\zeta) = 0$, $n \geq 0$. Hence $u(z) = 0$, $|z| > R$ and so, by uniqueness of analytic continuation, u vanishes identically on the unbounded component of $\mathbb{C} \setminus K$. On the other hand if $z \in \mathbb{C} \setminus \Omega$, $\partial^k u/\partial z^k = (-1)^k k! \int_{\mathbb{C}} (z - \zeta)^{-k-1} d\mu(\zeta) = 0$ since $(z - \zeta)^{-k-1} \in A(\Omega)$ for $z \in \mathbb{C} \setminus \Omega$. Consequently, $u \equiv 0$ on every component of $\mathbb{C} \setminus K$ which intersects $\Omega \setminus K$. But, by assumption, $\Omega \setminus K$ has no components which are relatively compact in Ω and so $u \equiv 0$ on $\mathbb{C} \setminus K$. □

Theorem A1.8. Let $f \in C^\infty(\Omega)$. Then there exists $u \in C^\infty(\Omega)$ such that

$$\partial u/\partial \bar{z} = f.$$

Proof. (Following Hörmander [1]). We already know from Theorem A1.5 that we can solve $\partial u/\partial \bar{z} = f$ if f has compact support. What we shall do is to write f as an infinite sum of C^∞ functions with compact support and then use Runge's theorem to modify the resulting infinite set of solutions so that they converge to a solution of $\partial u/\partial \bar{z} = f$.

For $n = 1,2,\ldots$, define $K_n = \{z \in \Omega : |z| \leq n \text{ and } d(z, \partial\Omega) \geq 1/n\}$. Then, the K_n are compact; $\Omega \setminus K_n$ has no components which are relatively compact in Ω; every compact subset of Ω is contained in some K_n. For $n = 1,2,\ldots$, choose $\theta_n \in C_c^\infty(\Omega, \mathbb{R})$ so that $\theta_n \equiv 1$ on a neighbourhood of K_n. Set $\phi_n = \theta_n - \theta_{n-1}$, $n > 1$. Clearly $\theta_n \equiv 0$ on a neighbourhood of K_{n-1} and $\Sigma\phi_n \equiv 1$ on Ω. Let $u_n \in C^\infty(\Omega)$ be a solution of $\partial u/\partial \bar{z} = \phi_n f$ given by Theorem A1.6. By the Runge Theorem, there exists $a_n \in A(\Omega)$ such that $|a_n - u_n| < 2^{-n}$ on K_{n-1}. Set $u = \sum_n (u_n - a_n)$ and note that the sum converges uniformly on any compact subset of Ω. For fixed N, the sum from N to ∞ consists of terms which are analytic on a neighbourhood of K_N and so defines an analytic function on the interior of K_N. Hence $u \in C^\infty(\Omega)$ and, differentiating term-by-term, $\partial u/\partial \bar{z} = f$. □

Remark. As we shall see later, Theorems A1.7 and A1.8 do not generalise to arbitrary domains in $\mathbb{C}^n$, $n > 1$.

Exercises.

1. Let Ω be a domain in $\mathbb{C}$. Suppose that $\omega \subset \Omega$ is an open neighbourhood of the compact subset K of Ω. For integers $p \geq 1$ and $u \in A(\Omega)$ let

$$|u|_p^\omega = \left(\int_\omega |u|^p dxdy\right)^{1/p} \quad (|u|_p^\omega \text{ may be infinite}).$$

Shown that, given s, $0 < s < d(K,\partial\Omega)$, there exists $C > 0$ such that for all $u \in A(\Omega)$ we have

$$\|\partial^j u/\partial z^j\|_K \leq \frac{Cj!}{s^j} |u|_p^\omega, \; j \geq 0.$$

(Hint: Choose $\phi \in C_c^\infty(\Omega)$ such that $\phi(z) = 1$ whenever $d(z,K) \leq s$ and apply Theorem A1.2 to ϕu^p).

2. Let Ω be a domain in $\mathbb{C}$ and for integers $p \geq 1$ let $L^p(\Omega) = \{f \in A(\Omega), |f|_p^\Omega < \infty\}$. Show that $L^p(\Omega)$ is a Banach space and, in case $p = 2$, a Hilbert space (Use the result of Q1. in combination with Corollary A1.4).

3. Show that if Ω is a simply connected domain in $\mathbb{C}$ and $f \in A(\Omega)$ then there is a sequence of polynomials converging to f uniformly on compact subsets of Ω. Show that if Ω is not simply connected, then there is a sequence of rational functions converging uniformly to f on compact subsets of Ω and that we may assume that the rational functions have all their poles outside Ω.

CHAPTER 2. FUNCTIONS OF SEVERAL COMPLEX VARIABLES

Introduction

In this Chapter we develop the basic theory of analytic functions of several complex variables with particular reference to the problem of extension of analytic functions. Section 1 is a straightforward generalisation of the one variable theory and contains such theorems as the power series representation of an analytic function. In section 2 we prove a Riemann removable singularities theorem for analytic functions of more than one variable. The material in section 3 is in sharp contrast to the onevariable theory and includes the theorem of Hartog's that implies, for example, that every analytic function on the punctured Euclidean disc in $\mathbb{C}^n$, $n > 1$, extends analytically to the whole disc. In section 4 we define domains of holomorphy and prove their equivalence with holomorphically convex domains. Section 5 is devoted to various pseudoconvexity properties that a domain of holomorphy possesses. In section 6 we discuss the Bergman kernel function of a domain and in section 7 we make a preliminary study of the Cousin problems.

§1. Elementary theory of analytic functions of several complex variables

In this section we shall make a preliminary study of analytic functions of more than one complex variable. Our development will follow that of §1 of Chapter 1 rather closely.

For $j = 1,\ldots,n$, we define the differential operators

$$\partial/\partial z_j = \tfrac{1}{2}(\partial/\partial x_j - i\partial/\partial y_j)$$

$$\partial/\partial \bar{z}_j = \tfrac{1}{2}(\partial/\partial x_j + i\partial/\partial y_j).$$

Just as in §4 of Chapter 1 we have a composite mapping formula for these operators.

Proposition 2.1.1. (Composite mapping formula) Let $U \subset \mathbb{C}^m$, $V \subset \mathbb{C}^n$ be open and $g \in C^1(U,\mathbb{C}^n)$, $f \in C^1(V)$. Suppose $g(U) \subset V$. Then

$$\partial(fg)/\partial z_j = \sum_{i=1}^{n} \left(\frac{\partial f}{\partial z_i}\frac{\partial g_i}{\partial w_j} + \frac{\partial f}{\partial \bar{z}_i}\overline{\frac{\partial g_i}{\partial \bar{w}_j}}\right)$$

$$\partial(fg)/\partial \bar{z}_j = \sum_{i=1}^{n} \left(\frac{\partial f}{\partial z_i}\frac{\partial g_i}{\partial \bar{w}_j} + \frac{\partial f}{\partial \bar{z}_i}\overline{\frac{\partial g_i}{\partial w_j}}\right),$$

where we denote coordinates on $\mathbb{C}^m$, $\mathbb{C}^n$ by $(z_1,\ldots,z_m)$, $(w_1,\ldots,w_n)$ respectively.

Proof. As in §5 of Chapter 1, we take real and imaginary parts and apply the usual composite mapping formula. We omit the tedious but elementary computations. □

For the remainder of this section Ω will always denote a domain in $\mathbb{C}^n$.

Definition 2.1.2. Let $f \in C^1(\Omega)$. We say that f is *analytic* or *holomorphic* on Ω if f is analytic in each variable separately. That is, if f satisfies the following system of first order partial differential equations:

$$\partial f/\partial \bar{z}_j = 0, \; 1 \le j \le n .$$

Remarks.

1. Let $f \in C^1(\Omega)$. The derivative of f at $z \in \Omega$, $Df(z): \mathbb{C}^n \to \mathbb{C}$, is an $\mathbb{R}$-linear map. It is easy to verify that f is analytic on Ω iff $Df(z)$ is $\mathbb{C}$-linear for all $z \in \Omega$. If f is analytic then the matrix of $Df(z)$, relative to the standard basis of $\mathbb{C}^n$, is $[\partial f/\partial z_1,\ldots,\partial f/\partial z_n]$.

2. If $f = (f_1,\ldots,f_m) \in C^1(\Omega,\mathbb{C}^m)$, we say that f is analytic if each f_i is analytic. It follows from the previous remark that $Df(z): \mathbb{C}^n \to \mathbb{C}^m$ is $\mathbb{C}$-linear with matrix $[\partial f_i/\partial z_j]$.

3. A consequence of Proposition 2.1.1 is that the composite of analytic maps is analytic.

4. Since the inverse of a $\mathbb{C}$-linear map is $\mathbb{C}$-linear, a holomorphic diffeomorphism is necessarily biholomorphic. For the same reason, the inverse function theorem, implicit function theorem and rank theorem all hold for analytic mappings (see Dieudonné [1] or Field [1]).

Notation. We denote the set of analytic functions on Ω by $A(\Omega)$.

As in the 1-variable theory, the main technical tool used in developing the elementary theory of analytic functions of several complex variables is an integral representation formula. First, however, some notational conventions.

Given $a = (a_1,\ldots,a_n) \in \mathbb{C}^n$ and $r_1,\ldots,r_n > 0$, we set

$$D(a;r_1,\ldots,r_n) = \prod_{i=1}^{n} D_{r_i}(a_i)$$

$$= \{z \in \mathbb{C}^n \colon |z_j - a_j| < r_j\}.$$

We call $D(a;r_1,\ldots,r_n)$ an open *polydisc* with centre a. We may similarly define closed polydiscs. In case $r_1 = \ldots = r_n = r$, we set $D(a;r_1,\ldots,r_n) = D(a;r)$ and call $D(a;r)$ the open polydisc of radius r, centre a. Note that $D(a;r)$ is just the open disc, radius r, centre a relative to the norm $|z| = \max_i |z_i|$ on $\mathbb{C}^n$.

Suppose $D = \prod_{i=1}^{n} D_i$ is a polydisc in $\mathbb{C}^n$. Thus each D_i will be a disc in $\mathbb{C}$. We let ∂D_j denote the boundary of D_j and define

$$\partial_0 D = \prod_{i=1}^{n} \partial D_j .$$

We call $\partial_0 D$ the *distinguished boundary* of D. We remark that $\partial_0 D$ is a proper subset of ∂D and is homeomorphic to a real n-dimensional torus.

Given $a = (a_1,\ldots,a_n) \in \mathbb{C}^n$ and $r > 0$, we set

$$E(z;r) = \{z \in \mathbb{C}^n \colon \sum_{i=1}^{n} |z_i - a_i|^2 < r^2\}.$$

We call $E(z;r)$ the open *Euclidean disc*, centre a, radius r. In case $z = 0$, we often abbreviate $E(z;r)$ to $E(r)$.

Both polydiscs and Euclidean discs will play an important role in the sequel.

Next we briefly review some multi-index notation. If $\mathbb{N}$ denotes the positive integers, including zero, and n is a strictly positive integer, $m \in \mathbb{N}^n$ is to be thought of as an n-tuple $(m_1,\ldots,m_n)$ of positive integers. Then $m!$ is shorthand for $m_1!\ldots m_n!$; $|m|$ for $|m_1+\ldots+m_n|$; ∂^m for $\partial^{m_1}/\partial z_1^{m_1}\ldots\partial^{m_n}/\partial z_n^{m_n}$; r^m for $r_1^{m_1}\ldots r_n^{m_n}$; z^m for $z_1^{m_1}\ldots z_n^{m_n}$; $f_m(z)$ for $f_{m_1\ldots m_n}(z)$.

Theorem 2.1.3. (Cauchy's integral formula for polydiscs). Let D be an open polydisc in $\mathbb{C}^n$ with centre a and suppose $f \in C^0(\bar{D})$ is analytic on D. Then for all $z = (z_1,\ldots,z_n) \in D$, we have

$$f(z) = (2\pi i)^{-n}\int_{\partial_0 D} f(\zeta_1,\ldots,\zeta_n)\prod_{i=1}^{n}(\zeta_i - z_i)^{-1}d\zeta_1\ldots d\zeta_n .$$

In particular, f is C^∞ and

$$\partial^m f(a) = (2\pi i)^{-n} m! \int_{\partial_0 D} f(\zeta_1,\ldots,\zeta_n)\prod_{i=1}^{n}(\zeta_i - a_i)^{-m_i-1} d\zeta_1\ldots d\zeta_n .$$

Proof. Repeated application of the Cauchy formula in 1-variable (Corollary A1.3) gives

$$f(z) = (2\pi i)^{-n}\int_{\partial D_1}\frac{d\zeta_1}{\zeta_1-z_1}\ldots\int_{\partial D_n}\frac{d\zeta_n}{\zeta_n - z_n} f(\zeta_1,\ldots,\zeta_n),\ z \in D .$$

For fixed z, the integrand is continuous on a compact domain of integration and so the integral formula follows from Fubini's theorem. The remaining statements are immediate by differentiation under the integral sign. □

Corollary 2.1.4. Every analytic function on an open subset of $\mathbb{C}^n$ is C^∞.

Proof. Immediate from Theorem 2.1.3 since infinite differentiability is a local property. □

Remarks.

1. The proof of Theroem 2.1.3 only uses the analyticity of f in each variable *separately* together with the continuity of f on $\bar{D}$. This observation, together with Corollary 2.1.4, proves Osgood's Lemma: If $f \in C^0(\Omega)$ and is separately analytic on Ω then f is analytic on Ω. Much less trivial is the amazing theorem of Hartogs which states that if f is

separately analytic, with *no* assumption of continuity on f, then f is analytic. This result is definitely false for real analytic functions as the simple example $f(x,y) = xy/(x^2 + y^2)$, $(x,y) \neq (0,0)$; $f(0,0) = 0$ shows. For proofs of Hartogs theorem we refer the reader to Hörmander [1] or R. Narasimhan [2].

2. The reader should note that we only integrate over the proper subset $\partial_0 D$ of ∂D in Cauchy's integral formula. Another, related, way of seeing the significance of $\partial_0 D$ is to observe that f must always take its maximum on $\partial_0 D$. Indeed, this is a simple consequence of the maximum principle for analytic functions of one variable ($\partial_0 D$ is actually the *Šilov boundary* of D. For general facts about Šilov boundaries see Fuks [1] or Gamelin[1]).

3. Although we shall not need them in the sequel, we wish to point out that there are several other integral representation formulae for analytic functions of more than one variable that are of considerable importance in some applications. We mention in particular the Bochner-Martinelli, Bergmann-Cauchy-Weil and Cauchy-Fantappié integral formulae. For the first two formulae we refer to Fuks [2] or Vladimirov [1] and for the third to Leray [1] or Aizenberg [1]. For the relations between these formulae and Cauchy's integral formula we refer to Harvey [1].

Theorem 2.1.5. Let $f \in A(\Omega)$ and $D \subset \Omega$ be a polydisc, centre a. Then

$$f(z) = \sum_m a_m (z-a)^m, \quad z \in D ,$$

where convergence is uniform on compact subsets of D and the coefficients a_m are given by $m!a_m = \partial^m f(a)$, $m \in \mathbb{N}^n$.

Proof. Let $D' \subset D$ be any open polydisc, centre a, such that $\bar{D}' \subset \Omega$ (that is, D' is relatively compact in Ω). By Theorem 2.1.3, we have

$$f(z) = (2\pi i)^{-n} \int_{\partial_0 D'} f(\zeta_1,\ldots,\zeta_n) \prod_{i=1}^n (\zeta_i - z_i)^{-1} d\zeta_1 \ldots d\zeta_n, \quad z \in D'.$$

Now if $z \in D'$, $\zeta \in \partial_0 D'$ we have

$$\prod_{i=1}^n (\zeta_i - z_i)^{-1} = \sum_m \frac{(z-a)^m}{(\zeta-a)^m} (\zeta - a)^{-1} ,$$

where $(\zeta - a)^{-1}$ is shorthand for $(\zeta_1 - a_1)^{-1} \ldots (\zeta_n - a_n)^{-1}$. The convergence of the sum on the right is uniform on compact subsets of D'. Multiply by $f(\zeta_1, \ldots, \zeta_n)$ and integrate term by term to obtain

$$\begin{aligned} f(z) &= (2\pi i)^{-n} \sum_m (z-a)^m \int_{\partial_0 D} \frac{f(\zeta_1, \ldots, \zeta_n)}{(\zeta - a)^m} (\zeta - a)^{-1} d\zeta_1 \ldots d\zeta_n \\ &= \sum_m a_m (z-a)^m. \end{aligned}$$

Differentiation of the power series and evaluation at a give the formulae for a_m. □

The proofs of the next four results parallel those of their one variable counterparts and we omit them.

Corollary 2.1.6. If $f \in A(\Omega)$, then $\partial^m f \in A(\Omega)$ for all $m \in \mathbb{N}^n$.

Corollary 2.1.7. (Cauchy's inequalities). If f is analytic on the polydisc $D = D(a; r_1, \ldots, r_n)$ and $|f| \le M$ on D, then

$$|\partial^m f(a)| \le M m! r^{-m}, \quad m \in \mathbb{N}^n.$$

Corollary 2.1.8. Let u_n, $n \ge 1$, be a sequence of analytic functions on Ω which converges uniformly on compact subsets of Ω to a function u. Then $u \in A(\Omega)$.

Corollary 2.1.9. (Montel's theorem). Let u_n, $n \ge 1$, be a sequence in $A(\Omega)$ which is uniformly bounded on every compact subset of Ω. Then there is a subsequence of the u_n which converges uniformly on compact subsets of Ω to a limit $u \in A(\Omega)$.

Examples.

1. Let $P(z) = \sum a_m z^m$, where the $a_m \in \mathbb{C}$ and only finitely many are non-zero. Then P is analytic on $\mathbb{C}^n$ and is called a *polynomial* (in n-variables).

2. Let $f, g \in A(\Omega)$ and suppose g is not identically zero and has zero set $Z(g)$. Then f/g defines an analytic function on $\Omega \setminus Z(g)$.

3. If $f \in A(\mathbb{C}^n)$, we call f an *entire* analytic function. Every polynomial is entire. So also are the Laplace transforms of functions

(or distributions) with compact support (see, for example, Hörmander [2] or Vladimirov [2]).

Theorem 2.1.10. (Laurent series). Suppose f is analytic in the annular region $A = \{z \in \mathbb{C}^n: r_j < |z_j| < R_j, j = 1,\ldots,n\}$. Then

$$f(z) = \sum_{|m|=-\infty}^{+\infty} a_m z^m, \quad z \in A ,$$

where the a_m are determined uniquely by f and A and convergence is uniform on compact subsets of A.

Proof. Exactly as for the 1-variable case using Cauchy's integral formula. Uniqueness of the coefficients a_m follows by induction on n. □

Proposition 2.1.11. (Uniqueness of analytic continuation). Let U, V be connected open subsets of $\mathbb{C}^n$ and suppose $U \cap V \neq \emptyset$. If $f \in A(U)$ and h is an analytic extension of f to $U \cup V$ then h is unique.

Proof. Exactly as for Proposition 1.1.6. □

Theorem 2.1.12. (Maximum Principle). Let $f \in A(\Omega)$ and suppose there exists $\zeta \in \Omega$ such that $|f(z)| \leq |f(\zeta)|$ for all $z \in \Omega$. Then f is constant.

Proof. Choose $r > 0$ so that the Euclidean disc $E = E(\zeta;r)$ is contained in Ω. If L denotes any (affine) complex line through ζ, the Maximum Principle for analytic functions of one complex variable applied to $f|(L \cap E)$ implies that f is constant on $L \cap E$. This holds for all complex lines through ζ and so f is constant on E. Uniqueness of analytic continuation now implies that f is constant on Ω. □

Finally, another simple application of one-variable theory.

Theorem 2.1.13. (Open Mapping Theorem). Let $f \in A(\Omega)$ and suppose that f is not constant. Then f is an open mapping. That is, f maps open subsets of Ω to open subsets of $\mathbb{C}$.

Proof. It is enough to prove that if $E \subset \Omega$ is any (Euclidean) open disc neighbourhood of $\zeta \in \Omega$, then $f(\zeta)$ is an interior point of $f(E)$. Since f is not constant, it follows by uniqueness of analytic continuation that $f|E \cap L$ is not constant on every complex line L through (compare the proof of Theorem 2.1.1). But if $f|E \cap L$ is not constant, $f(E \cap L)$ is a neighbourhood of $f(\zeta)$ by the well-known one-variable result (Exercise 4, §1, Chapter 1). □

Exercises.

1. Let Ω be a domain in $\mathbb{C}^n$. We say Ω is a *Reinhardt domain* if $(z_1,\dots,z_n) \in \Omega$ implies $(e^{i\theta_1}z_1,\dots,e^{i\theta_n}z_n) \in \Omega$ for all $\theta_1,\dots,\theta_n \in \mathbb{R}$. Show that if Ω is a Reinhardt domain containing the origin and $f \in A(\Omega)$ then

$$f(z) = \sum a_m z^m, \; z \in \Omega,$$

where convergence is uniform on compact subsets of Ω. (Hint: Given $\varepsilon > 0$, let $\Omega_\varepsilon = \{z \in \Omega\colon d(z,\partial\Omega) > \varepsilon|z|\}$ and Ω'_ε denote the connected component of Ω_ε containing the origin. For $z \in \Omega'_\varepsilon$, define $g(z) = (2\pi i)^{-n} \int_{\partial_0 D} f(t_1z_1,\dots,t_nz_n)(t-1)^{-1}dt_1\dots dt_n$, where $(t-1)^{-1}$ is shorthand for $(t_1-1)^{-1}\dots(t_n-1)^{-1}$ and $D = D(0;1+\varepsilon)$. Note that if $z \in \Omega'_\varepsilon$, then $(1+\varepsilon)z \in \Omega$. Certainly g is analytic in Ω'_ε. Show that $f(z) = g(z)$ for $|z|$ sufficiently small. Expand $(t-1)^{-1}$ as a Laurent series in t and integrate term by term to obtain a power series for f. Then let $\varepsilon \to 0$ and note that $\Omega = \bigcup_{\varepsilon>0} \Omega'_\varepsilon$).

Deduce that if Ω is a Reinhardt domain containing the origin then the polynomials are dense in $A(\Omega)$ (topology of uniform convergence on compact subsets). In particular, the polynomials are dense in $A(E(0;r))$, $r > 0$.

2. Let Ω be a domain in $\mathbb{C}^n$ and $\omega \subset \Omega$ be an open neighbourhood of the compact subset K of Ω. For integers $p \geq 1$ and $u \in A(\Omega)$ let

$$|u|_p^\omega = \left(\int_\omega |u|^p d\lambda\right)^{1/p},$$

where $d\lambda$ denotes Lebesgue measure on $\mathbb{C}^n$ ($|u|_p^\omega$ may be infinite). Show that for all $m \in \mathbb{N}^n$ there exist constants C_m such that

$$\sup_{z\in K} |\partial^m u(z)| = \|\partial^m u\|_K \leq C_m |u|_p^\omega, \; u \in A(\Omega).$$

Deduce that the spaces $L^p(\Omega) = \{f \in A(\Omega): |f|_p^\Omega < \infty\}$ are Banach spaces and that $L^2(\Omega)$ is a Hilbert space. (Hint for the first part: Repeated application of Corollary A1.4, see also Exercises 1,2 in the Appendix to Chapter 1).

§2. Removable Singularities

The Riemann removable singularities theorem for functions of one complex variable (§2, Chapter 1) implies that if X is a discrete subset of $\Omega \subset \mathbb{C}$ then every analytic function on $\Omega \setminus X$ which is locally bounded on Ω has a unique analytic extension to Ω. In this section we generalise this result to analytic functions of more than one variable.

Definition 2.2.1. Let X be a subset of the domain Ω in $\mathbb{C}^n$. We say that X is an *analytic subset* of Ω if for every $z \in \Omega$ there exists an open neighbourhood U of z in Ω and analytic map $f: U \to \mathbb{C}^p$ such that $X \cap U = f^{-1}(0)$ (p may depend on z).

We shall undertake a more systematic study of analytic sets in Chapters 3 and 4 and for the present we remark only that our definition implies that if X is an analytic subset of Ω then X is a closed subset of Ω.

Theorem 2.2.2. (Riemann removable singularities theorem). Let X be a proper analytic subset of the domain Ω in $\mathbb{C}^n$ and $f \in A(\Omega \setminus X)$. Suppose that for every $x \in \Omega$, there exists an open neighbourhood U of x in Ω such that $f|U \setminus X$ is bounded. Then f has a unique analytic extension to Ω.

Proof. It is clearly sufficient to show that we can find an open neighbourhood D of every $z_0 \in \Omega$ such that $f|D \setminus X$ extends uniquely to D. The case $z_0 \notin X$ being trivial we shall suppose $z_0 \in X$. Choose an open connected neighbourhood U of z_0 so that $X \cap U = g^{-1}(0)$ for some $g \in A(U,\mathbb{C}^p)$. Writing $g = (g_1,\ldots,g_p)$, we may suppose $h = g_1 \not\equiv 0$. Since h cannot be identically zero on every affine complex line through z_0, we may make an affine linear change of coordinates and suppose that $z_0 = 0$ and $h(0,\ldots,0,z_n) \not\equiv 0$ on a neighbourhood of $z_n = 0$. Since $z_n = 0$ is an isolated zero of $h(0,\ldots,0,z_n)$, there exists $\delta > 0$ such that $h(0,\ldots,0,z_n) \neq 0$ for $0 < |z_n| \leq \delta$. Denote the variable $(z_1,\ldots,z_{n-1}) \in \mathbb{C}^{n-1}$ by z' and set $|z'| = \max_i |z_i|$. By the continuity of h we may choose $r > 0$

so that $h(z',z_n) \neq 0$ for $|z'| \leq r$ and $|z_n| = \delta$. Let D denote the polydisc $\{(z',z_n): |z'| < r, |z_n| < \delta\}$. We define

$$g(z',z_n) = (2\pi i)^{-1} \int_{|\zeta|=\delta} \frac{f(z',\zeta)}{\zeta - z_n} d\zeta, \ (z',z_n) \in D.$$

Since $f(z',\zeta)$ is holomorphic in z' for $|z'| \leq r$ and $|\zeta| = \delta$, $g \in A(D)$.

For fixed z', $|z'| < r$, the function $f(z',\zeta)$ extends by the one variable Riemann removable singularities theorem to the disc $|\zeta| < \delta$. Hence, by the Cauchy integral formula, $g(z) = f(z)$ if $z \in D \setminus X$. □

Corollary 2.2.3. Let X be a proper analytic subset of the domain Ω in $\mathbb{C}^n$. Then $\Omega \setminus X$ is an open, connected and dense subset of Ω.

Proof. If $\Omega \setminus X$ were not connected we could define $f \in A(\Omega \setminus X)$ by taking $f \equiv 0$ on one connected component of $\Omega \setminus X$ and $f \equiv 1$ on all the other connected components. The Riemann removable singularities theorem then implies that f extends uniquely to Ω and we obtain a contradiction by the uniqueness of analytic continuation. We leave the remaining assertion as an exercise. □

Next we state an important removable singularities theorem due to Rado.

Theorem 2.2.4. (Rado's Theorem). Let Ω be a domain in $\mathbb{C}^n$ and $f \in C^0(\Omega)$. Suppose that f is analytic on $\Omega \setminus f^{-1}(0)$. Then f is analytic on Ω.

Proof. One possible proof of this theorem depends on a reduction to the one-variable case using Hartog's theorem (Remark 1, §1). The one-variable proof uses the theory of harmonic and subharmonic functions. We refer the reader to R. Narasimhan [2] for more details. An alternative elementary proof which avoids the use of Hartogs theorem and makes minimal use of subharmonic functions is given in Whitney [1]. □

Remark. We shall not make any use of Rado's theorem in these notes. Rado's theorem does, however, have important applications, notably to the theory of biholomorphic maps and automorphisms. By way of example, we mention Osgood's theorem: If $f: \Omega \subset \mathbb{C}^n \to \mathbb{C}^n$ is an injective holomorphic map then $f(\Omega)$ is open and f is biholomorphic onto its image (The example $f(x) = x^3$, shows that this result is not true for real analytic maps). For a proof of Osgood's theorem and further examples and references, we refer to R. Narasimhan [2].

We conclude this section by stating an important result on the singularities of analytic functions due to Hartog's. Suppose that Z is a subset of the domain Ω in $\mathbb{C}^n$ and f is holomorphic on $\Omega \setminus Z$. We say that f is *singular* at $z \in Z$ if there is no holomorphic function defined on a neighbourhood U of z in Ω whose restriction to $U \setminus Z$ is $f|U \setminus Z$.

Theorem 2.2.5. (Hartogs). Let Ω be a domain in $\mathbb{C}^{n-1}$ and suppose $\phi: \Omega \to D_R(0)$ is a map. Let $\Sigma = \{(z,t) \in \Omega \times D_R(0): \phi(z) = t\}$ and suppose $f \in A(\Omega \times D_R(0) \setminus \Sigma)$. If every point of Σ is singular for f then ϕ is holomorphic. In particular, Σ is an analytic subset of $\Omega \times D_R(0)$.

Proof. A proof, depending on the theory of subharmonic functions, may be found in R. Narasimhan [2]. □

As a straightforward corollary of Theorem 2.2.5 we have

Corollary 2.2.6. Let Σ be a (real) k-dimensional submanifold of the domain Ω in $\mathbb{C}^n$. Then

1. If $k < 2n-2$, every analytic function on $\Omega \setminus \Sigma$ extends to Ω.

2. If $k = 2n-2$ and there do not exist any points $x \in \Sigma$ which have a neighbourhood U such that $U \cap \Sigma$ is an analytic subset of U, then every analytic function on $\Omega \setminus \Sigma$ extends to Ω.

Exercises.

1. Let A be a subset of the domain Ω in $\mathbb{C}^n$. We say that A is *thin* if for every $z \in \Omega$, we may find a polydisc $D(z;r)$ and $f \in A(D(z;r))$, not identically zero, such that $A \cap U \subset f^{-1}(0)$. Show that if A is a thin subset of Ω and f is analytic on $\Omega \setminus A$ and locally bounded on Ω, then f extends uniquely to an analytic function on Ω.

2. Prove part 1 of Corollary 2.2.6 directly without using Theorem 2.2.5 (Use Cauchy's integral formula).

§3. Extension of analytic functions

Thus far our development of the theory of analytic functions of more than one complex variable has paralleled the one-variable theory very closely. In this section, we start investigating phenomena peculiar to analytic functions of two or more complex variables.

Examples.

1. Let f be an analytic function on $\mathbb{C}^n \setminus \{0\}$, $n > 1$. Then f extends analytically to $\mathbb{C}^n$. Indeed, for $(z_1,\ldots,z_n) \in \mathbb{C}^{n-1} \times D_1(0)$ we may define

$$F(z_1,\ldots,z_n) = (2\pi i)^{-1} \int_{|\zeta|=1} f(z_1,\ldots,z_{n-1},\zeta)/(\zeta - z_n)d\zeta .$$

Certainly $F \in A(\mathbb{C}^{n-1} \times D_1(0))$ and by Cauchy's integral formula $F(z_1,\ldots,z_n) = f(z_1,\ldots,z_n)$ provided $(z_1,\ldots,z_{n-1}) \neq 0$. Hence, by uniqueness of analytic continuation, $F = f$ on $(\mathbb{C}^{n-1} \times D_1(0)) \setminus \{0\}$. Setting $F = f$ outside $\mathbb{C}^{n-1} \times D_1(0)$ we see that F is the required analytic extension of f to $\mathbb{C}^n$. This result is definitely false if $n = 1$ (take $f(z) = z^{-1}$!) and showsthat Corollary 1.3.5 fails for arbitrary domains in $\mathbb{C}^n$, $n > 1$. That is, an arbitrary open subset of $\mathbb{C}^n$ will not generally be the domain of existence of an analytic function if $n > 1$.

2. Let D be an open polydisc in $\mathbb{C}^n$ and suppose that K is a compact subset of D such that $D \setminus K$ is connected. Then every analytic function on $D \setminus K$ extends uniquely to an analytic function on D. The proof uses the same idea as in example 1 and we leave details to the reader. Notice that in this example K may have interior points. Later in this section we prove the far more general theorem of Hartogs: If K is a compact subset of $\Omega \subset \mathbb{C}^n$, $n > 1$, and $\Omega \setminus K$ is connected, then every analytic function on $\Omega \setminus K$ extends uniquely to Ω.

3. Let $D = \{(z_1,\ldots,z_n): |z_i| < 1,\ 1 \le i \le n\}$ and

$$P = \{(z_1,\ldots,z_n) \in D: |z_i| < \tfrac{1}{2},\ 1 \le i \le n-1 \text{ or } |z_n| > \tfrac{1}{2}\}.$$

The real parts of D and P are pictured in Figure 2.

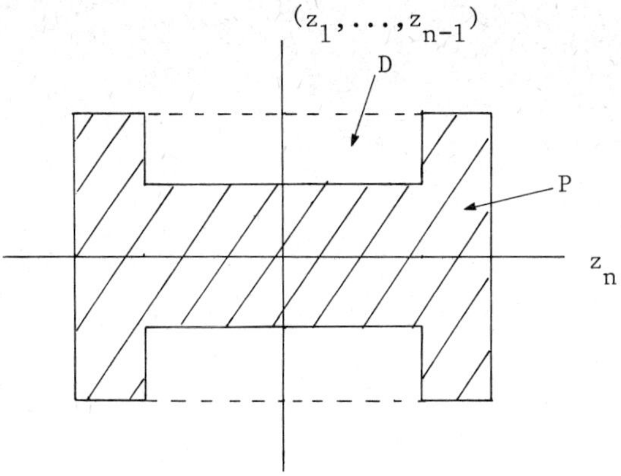

Figure 2.

We claim that every analytic function on P extends analytically to D. Our proof is similar to that of example 1. Suppose $f \in A(P)$, we define

$$F(z_1,\dots,z_n) \;=\; (2\pi i)^{-1} \int_{|\zeta|=3/4} f(z_1,\dots,z_{n-1},\zeta)/(\zeta - z_n)d\zeta,$$

where $|z_i| < 1$, $1 \le i \le n-1$ and $|z_n| < 3/4$. Certainly F is analytic and if $|z_1|,\dots,|z_{n-1}| < \frac{1}{2}$ we see that $F(z_1,\dots,z_n) = f(z_1,\dots,z_n)$ by Cauchy's integral formula. It now follows by uniqueness of analytic continuation that $F = f$ wherever both are defined and setting $F = f$ for $1 > |z_n| \ge 3/4$ we see that F is the required analytic extension of f to D. The pair (D,P) is called a *Euclidean Hartogs figure*.

4. Suppose (D,P) is as in example 3 and $g: D \to \mathbb{C}^n$ is biholomorphic onto the image of g. Set $\tilde{D} = g(D)$, $\tilde{P} = g(P)$. We claim that every analytic function on $\tilde{P}$ extends uniquely to an analytic function on $\tilde{D}$. Indeed, if $f \in A(\tilde{P})$, $gf \in A(P)$. Now fg extends by example 3 to $F \in A(D)$. Clearly $Fg^{-1} \in A(\tilde{D})$ is the required analytic extension of f to $\tilde{D}$. The pair $(\tilde{D},\tilde{P})$ is called a *generalised Hartogs figure*. As we shall see later in this chapter we can use generalised Hartogs figures as a test to determine whether or not every analytic function on a given domain can be extended to a larger domain.

The remainder of this section is devoted to a proving the Theorem of Hartogs referred to in example 2 above. The proof follows Ehrenpreis [1] and Hörmander [1] and uses an existence theorem for the $\partial/\partial\bar{z}_j$ operators.

Suppose $f_j \in C^\infty(\mathbb{C}^n)$, $j = 1,\dots,n$ and that we wish to solve the system of partial differential equations

$$\partial u/\partial\bar{z}_j = f_j, \; 1 \le j \le n.$$

We first remark that this system is overdetermined in that for solvability the additional conditions

$$\partial f_i/\partial\bar{z}_j = \partial f_j/\partial\bar{z}_i, \; 1 \le i, j \le n$$

obviously have to be satisfied.

Theorem 2.3.1. Let $f_j \in C_c^\infty(\mathbb{C}^n)$, $1 \le j \le n$ and suppose that $\partial f_i/\partial\bar{z}_j = \partial f_j/\partial\bar{z}_i$, $1 \le i, j \le n$. Then the system of equations

$$\partial u/\partial\bar{z}_j = f_j$$

has a solution $u \in C_c^\infty(\mathbb{C}^n)$ provided that $n > 1$.

Proof. Define

$$u(z) = (2\pi i)^{-1} \int_{\mathbb{C}} f_1(\zeta, z_2, \dots, z_n)/(\zeta - z_1)\, d\zeta d\bar{\zeta} .$$

Changing variables we see that

$$u(z) = -(2\pi i)^{-1} \int_{\mathbb{C}} f_1(z_1 - \zeta, z_2, \dots, z_n)\zeta^{-1} d\zeta d\bar{\zeta}$$

and so $u \in C^\infty(\mathbb{C}^n)$. Furthermore, since f_1 has compact support, $u(z) = 0$ provided that $|z_2| + \dots + |z_n|$ is sufficiently large.

Theorem A1.6 implies that $\partial u/\partial\bar{z}_1 = f_1$. Differentiating under the integral sign with respect to $\bar{z}_j$ and using the relation $\partial f_j/\partial\bar{z}_1 = \partial f_1/\partial\bar{z}_j$, we obtain

$$\begin{aligned}\partial u/\partial\bar{z}_j &= (2\pi i)^{-1} \int_{\mathbb{C}} (\zeta - z_1)^{-1} \partial f_j/\partial\bar{\zeta}(\zeta, z_2, \dots, z_n) d\zeta d\bar{\zeta} \\ &= f_j, \text{ by Theorem A1.2.}\end{aligned}$$

Hence u is a solution of the given system of equations. Now let K denote the union of the supports of the f_j. Then $u \in A(\mathbb{C}^n \setminus K)$ and u

is zero for $|z_2| + \ldots + |z_n|$ sufficiently large. Hence, by uniqueness of analytic continuation, u is zero on the unbounded component of $\mathbb{C}^n \setminus K$ and so u has compact support. □

Remarks.

1. As we remarked in the appendix to Chapter 1, Theorem 2.3.1 is false for n = 1.

2. Later, in Chapter 5, we shall rewrite the system of partial differential equations occuring in Theorem 2.3.1 in the framework of differential forms.

Theorem 2.3.2. (Hartog's theorem). Let Ω be a domain in $\mathbb{C}^n$, $n > 1$, and K be a compact subset of Ω such that $\Omega \setminus K$ is connected. Then every analytic function on $\Omega \setminus K$ extends uniquely to an analytic function on Ω.

Proof. Choose $\theta \in C_c^\infty(\Omega)$ so that $\theta \equiv 1$ on K. Define $f_0 \in C^\infty(\Omega)$ by setting

$$f_0|K = 0; \quad f_0|\Omega \setminus K = (1-\theta)f .$$

We shall construct $v \in C^\infty(\mathbb{C}^n)$ so that $f_0 + v$ is the required continuation of f. First notice that $f_0 + v$ will be analytic iff v satisfies

$$\begin{aligned} \partial v/\partial \bar{z}_j &= -\partial f_0/\partial \bar{z}_j, \; j = 1,\ldots,n \\ &= -\partial\theta/\partial\bar{z}_j f . \end{aligned}$$

Now $-\partial\theta/\partial\bar{z}_j\, f \in C_c^\infty(\mathbb{C}^n)$ and so we may apply Theorem 2.3.1 to find $v \in C_c^\infty(\mathbb{C}^n)$ such that $f_0 + v \in A(\Omega)$. Now observe that since v has compact support, v vanishes on the unbounded component of the complement of the support of θ - uniqueness of analytic continuation. Since $\text{supp}(\theta) \subset \Omega$, there exists an open set in $\Omega \setminus K$ where $v = 0$ and so $f = f_0$. Since $\Omega \setminus K$ is connected, $f = f_0 + v$ on $\Omega \setminus K$ and so $f_0 + v$ is the required analytic extension of f. □

Remark. An alternative proof of Hartog's theorem, based on the Bochner-Martinelli integral formula, may be found in Bochner [1] . See also Harvey [2; page 355].

Corollary 2.3.3. Let Ω be a domain in $\mathbb{C}^n$, $n > 1$, and $f \in A(\Omega)$. Then the zero set $Z(f)$ of f is never a compact subset of Ω.

Proof. If $Z(f)$ were compact, $1/f \in A(\Omega \setminus Z(f))$ would extend analytically to Ω by Theorem 2.3.2. □

The corollary empahsises an important difference between the one- and several-variable theory of analytic functions. The zero set of an analytic function of more than one variable always propagates to the boundary of the domain on which the function is defined. It is this essential non-compactness that makes the study of the zero and pole sets so much harder than for functions of one complex variable.

Exercise. Let Ω, ω be domains in $\mathbb{C}^n$, $\mathbb{C}$ respectively. A map $f: \Omega \to \omega$ is said to be *proper* if $f^{-1}(K)$ is compact whenever $K \subset \omega$ is compact. Prove that if $n > 1$, there are no proper holomorphic maps of Ω into ω (Hint: Let $z_0 \in f(\Omega)$ and consider $g(z) = (f(z) - z_0)^{-1} \in A(\Omega \setminus f^{-1}(z_0)))$.

§4. Domains of Holomorphy

In §3 we gave a large class of examples of domains in $\mathbb{C}^n$, $n > 1$, that possessed the property that every analytic function on them extended to a larger domain. In this section we wish to look at domains where this phenomenon does not happen and the generalisation of Corollary 1.3.5 is true.

Throughout this section Ω will denote a domain in $\mathbb{C}^n$.

Definition 2.4.1. A domain in $\mathbb{C}^n$ is called a *domain of holomorphy* if whenever we are given open connected subsets $U \subset V \subset \mathbb{C}^n$, with $U \subset \Omega$, such that $f|U$ extends analytically to V for all $f \in A(\Omega)$, then $V \subset \Omega$.

For Ω to be a domain of holomorphy we require that we cannot extend every analytic function on Ω locally across any point of the boundary of Ω. We frame the definition in this rather complicated fashion to exclude the following type of domain:

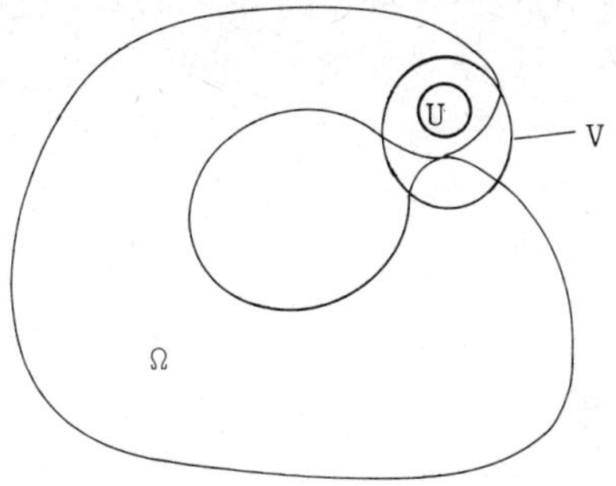

Figure 3.

Referring to the figure, suppose that $f|U$ extends to $F \in A(V)$ for all $f \in A(\Omega)$. Then Ω will not be a domain of holomorphy. Notice though that $F|\Omega \cap V$ need not equal $f|\Omega \cap V$ - we give an explicit example below.

Examples.

1. $\mathbb{C}^n$ is a domain of holomorphy, $n \geq 1$.

2. If K is a compact subset of $\Omega \subset \mathbb{C}^n$, $n > 1$, such that $\Omega \setminus K$ is connected then $\Omega \setminus K$ is not a domain of holomorphy (Theorem 2.3.2).

3. Every domain Ω in $\mathbb{C}$ is a domain of holomorphy. This is trivial: Given $a \in \partial\Omega$, define $f(z) = (z-a)^{-1} \in A(\Omega)$ and observe that f cannot extend to any open neighbourhood of a.

4. A necessary condition for Ω to be a domain of holomorphis is that if $(\tilde{D},\tilde{P})$ is any generalised Hartog's figure (Example 4, §3) with $\tilde{P} \subset \Omega$, then $\tilde{D} \subset \Omega$. Actually this condition turns out to be sufficient (see the discussion in §5).

5. Given open domains $U_i \subset \mathbb{C}$, $1 \leq i \leq n$, $U = \coprod_{i=1}^{n} U_i$ is a domain of holomorphy. Indeed, $\partial U = \cup_i U_1 \times \ldots \times \partial U_i \times \ldots \times U_n$ and so if $a = (a_1,\ldots,a_n) \in \partial U$, we must have at least one $a_i \in \partial U_i$. Now define $f(z_1,\ldots,z_n) = (z_i - a_i)^{-1}$. Clearly f does not extend to any open neighbourhood of a. Alternatively, choose $f_i \in A(U_i)$ satisfying the conditions of Corollary 1.3.5. Define $f(z_1,\ldots,z_n) = \sum_{i=1}^{n} f_i(z_i)$. In this case we see that U is the "domain of existence" for f.

6. Let $V = \{z \in D(0;1): z_1 = \dots z_{n-1} = 0, \operatorname{Im}(z_n) = 0, \operatorname{Re}(z_n) \geq 0\}$ and set $\Omega = D(0;1) \setminus V$. Every analytic function on Ω extends analytically to $D(0;1)$ (Exercise 2, §2 or use Laurent series at 0). On the other hand Ω is homeomorphic to $D(0;1)$ and so we see that the property of being a domain of holomorphy is not a topological invariant.

7. We now present an example to show that the phenomenon alluded to after definition 2.4.1 can occur. Our example is in $\mathbb{C}^2$, though it is easily generalised to $\mathbb{C}^n$, $n > 2$. Let $A \subset \mathbb{C}^2$ be the product of the "cut" annulus $\{z \in \mathbb{C}: \frac{1}{2} < |z-2| < 3/2\} \setminus \{z \in \mathbb{C}: |z| = \frac{1}{2} \text{ and } \arg(z) \in (0,\pi)\}$ with the disc $|w| < \frac{1}{4}$. We define Ω to be the union of A with $P = \{(z,w) \in D(0;1): |z| < \frac{1}{2} \text{ or } 1 > |w| > \frac{1}{2}\}$. Let $D = D(0;1)$.

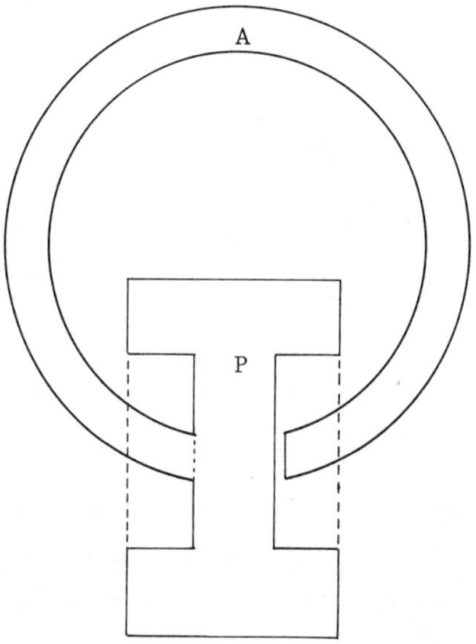

Figure 4.

Since (D,P) is a Euclidean Hartog's figure, $f|P$ extends analytically to $F \in A(D)$ for all $f \in A(\Omega)$. But in general $F|D \cap \Omega \neq f|D \cap \Omega$ as is seen by taking $f(z,w) = \sqrt{(z-2)}$.

Example 5 suggests that a domain of holomorphy might be the domain of existence of an analytic function (note that the converse is trivially true). Much of the remainder of this section will be devoted to proving that every domain of holomorphy is the domain of existence of some analytic function. In the course of our proof we shall derive other important characterizations of domains of holomorphy.

Example 3 and 5 also suggest that if Ω is a domain of holomorphy, $z \in \partial\Omega$ and $(z_n) \subset \Omega$ converges to z, then there exists $f \in A(\Omega)$ which is unbounded on the sequence (z_n). We make a formal definition.

Definition 2.4.2. We say that the domain Ω possesses property (S) if given any sequence $(z_n) \subset \Omega$ which converges to a point $z \in \partial\Omega$, there exists $f \in A(\Omega)$ which is unbounded on the sequence (z_n).

Lemma 2.4.3. If the domain Ω possesses property (S), Ω is a domain of holomorphy.

Proof. Suppose Ω is not a domain of holomorphy. Then there exists $z \in \partial\Omega$ and open connected sets $U \subset V \subset \mathbb{C}^n$ such that $U \subset \Omega$, $z \in V$ and $f|U$ extends to $F \in A(V)$ for every $f \in A(\Omega)$. Now let $(z_n) \subset \Omega \cap V$ converge to z. We see immediately that F is bounded on (z_n). Therefore f is bounded on (z_n) for all $f \in A(\Omega)$ and Ω does not possess property (S). □

Examples.

7. The Euclidean disc $E = \{z \in \mathbb{C}^n : \sum_{i=1}^{n} |z_i|^2 < r^2\}$ is a domain of holomorphy. We show E possesses property (S). Suppose $a = (a_1,\ldots,a_n) \in \partial E$. Define $f(z) = (r^2 - \langle z,a\rangle)^{-1}$, where $\langle\ ,\ \rangle$ denotes the standard Hermitian inner product on $\mathbb{C}^n$. Clearly $f \in A(E)$ and f is not bounded on any sequence of points of E converging to a.

8. Let $f_j \in A(\mathbb{C}^n)$, $1 \le j \le q$. The *analytic polyhedron* $P = \{z \in \mathbb{C}^n : |f_j(z)| < 1,\ 1 \le j \le q\}$ possesses property (S) and is a domain of holomorphy. Indeed, if $a \in \partial P$, $|f_i(a)| = 1$ for some i. Now define $F(z) = (f_i(z) - f_i(a))^{-1} \in A(P)$. Clearly F is not bounded on any sequence of points of P converging to a.

It turns out that in some ways domains of holomorphy are the complex analogue of convex sets. The next definition reflects these convexity properties particularly well. Before giving the definition we recall that if K is a compact subset of $\Omega \subset \mathbb{C}^n$ and $f: \Omega \to \mathbb{C}$, then $\|f\|_K$ is, by definition, $\sup_{z\in K} |f(z)|$.

Definition 2.4.4. A domain Ω is said to be *holomorphically convex* if given any compact subset $K \subset \Omega$ the set

$$\hat{K} = \{z \in \Omega: |f(z)| \leq \|f\|_K \text{ for all } f \in A(\Omega)\}$$

is compact.

Remark. We call $\hat{K}$ the $A(\Omega)$-hull of K. If $K = \hat{K}$, we say K is $A(\Omega)$-convex.

Examples.

9. Every domain Ω in $\mathbb{C}$ is holomorphically convex. In fact it is easy to see that $d(K,\partial\Omega) = d(\hat{K},\partial\Omega)$ for every compact subset K of Ω: Just consider functions $f(z) = (z - \zeta)^{-1} \in A(\Omega)$, where $\zeta \in \partial\Omega$.

10. The domain $\mathbb{C}^n \setminus \{0\}$, $n > 1$, is not holomorphically convex. Indeed suppose $K = \{z \in \mathbb{C}^n: |z| = 1\}$ (the boundary of the unit polydisc). Then $\hat{K} = \{z: 0 < |z| \leq 1\}$ (the punctured unit polydisc). To see this we notice that by Theorem 2.3.2 any $f \in A(\mathbb{C}^n \setminus \{0\})$ extends to $F \in A(\mathbb{C}^n)$. By the Maximum Principle (Theorem 2.1.12), the maximum value of F on $\{z: |z| \leq 1\}$ is taken on its boundary, K. Since F is an extension of f the same is true for f. This implies $\hat{K} \supseteq \{z: 0 < |z| \leq 1\}$. The reverse inclusion is trivial. Notice that if $n = 1$, $\hat{K} = K$ as is seen by taking $f(z) = z^{-1}$.

Before stating the next lemma we wish to review a few elementary facts about convex subsets of $\mathbb{R}^n$. Suppose $X \subset \mathbb{R}^n$. We say that X is *convex* if, given $x,y \in X$, $tx + (1-t)y \in X$ for all $t \in [0,1]$. Now suppose X is bounded. The (closed) *convex hull*, X_c, of X is defined to be

$$X_c = \{u \in \mathbb{R}^n: \phi(u) \leq \sup_{x\in X} \phi(x) \text{ for all } \phi \in \mathbb{R}^{n'}\}.$$

The following properties of the convex hull are easily verified:

X_c is closed, convex and bounded

$X_{cc} = X_c$

$X_c \supset X$ and X_c is the smallest closed convex set containing X

$X_c = X$ if and only if X is closed and convex.

Lemma 2.4.5. Let K be a compact subset of the domain Ω in $\mathbb{C}^n$. Then

1. $\hat{K} \supset K$.

2. $\hat{K}$ is a closed subset of Ω.

3. $\hat{\hat{K}} = \hat{K}$.

4. $\hat{K} \subset K_c$. In particular, $\hat{K}$ is bounded.

Proof. 1, 2 and 3 are trivial. Let us prove 4.

$$K_c = \{u \in \mathbb{C}^n: \phi(u) \le \sup_{z \in K} \phi(z) \text{ for all } \phi \in L_{\mathbb{R}}(\mathbb{C}^n, \mathbb{R})\} .$$

Given $\phi \in L_{\mathbb{R}}(\mathbb{C}^n, \mathbb{R})$, we may write

$$\phi(z_1,\ldots,z_n) = \sum_{i=1}^{n} a_i z_i + \sum_{i=1}^{n} b_i \bar{z}_i$$

where $a_i, b_i \in \mathbb{C}$. Since ϕ is real valued we must have $b_i = \bar{a}_i$, $i = 1,\ldots,n$, and so

$$\phi(z_1,\ldots,z_n) = 2\mathrm{Re} \sum_{i=1}^{n} a_i z_i .$$

Define $f(z) = \exp(2 \sum_{i=1}^{n} a_i z_i) \in A(\Omega)$. If $z \in \hat{K}$,

$$|f(z)| \le \|f\|_K .$$

That is,

$$|\exp \phi(z)| \le \|\exp \phi\|_K, \text{ since } |f(z)| = |\exp \phi(z)| .$$

This implies $\phi(z) \le \sup_K \phi$, since ϕ is real. Hence $z \in K_c$. □

Remark. The reader should note the close analogy between properties of the $A(\Omega)$-hull and the convex hull. It should also be noted that an open subset Ω of $\mathbb{R}^n$ is convex if and only if K_c is a compact subset of Ω for every compact subset K of Ω. Furthermore, convexivity is a local property of the boundary of Ω (Every disc centered on the boundary of Ω meets Ω in a convex set if Ω is convex and this property characterises the convexivity of Ω). We shall see in §5 that we can also formulate holomorphic convexivity in terms of local properties of the boundary.

Theorem 2.4.6. A domain Ω in $\mathbb{C}^n$ is holomorphically convex if and only if it possesses property (S). In particular, if Ω is holomorphically convex it is a domain of holomorphy.

Proof. We start by showing that (S) implies holomorphic convexivity. Suppose Ω possesses property (S). If Ω is not holomorphicall convex there exists a compact subset K of Ω such that $\hat{K}$ is not compact. Since $\hat{K}$ is bounded we can therefore find a sequence $(z_n) \subset \hat{K}$ which converges to some point of $\partial\Omega$. Now $|f(z_n)| \leq \|f\|_K$ for all $f \in A(\Omega)$ by definition of $\hat{K}$. Hence every $f \in A(\Omega)$ must be bounded on (z_n). This is contrary to our assumption that Ω possesses property (S) and so Ω is holomorphically convex.

Suppose Ω is holomorphically convex. Let $(z_n) \subset \Omega$ be a sequence converging to some point of $\partial\Omega$. It is sufficient to construct $f \in A(\Omega)$ which is unbounded on (z_n).

First we choose a sequence (K_n) of compact subsets of Ω satisfying

1. $K_n \subset K_{n+1}$, $n \geq 1$.

2. $\bigcup_n K_n = \Omega$.

3. $K_n = \hat{K}_n$, $n \geq 1$.

4. Every point of Ω is an interior point of some K_n.

To see that such sequences exist, we first construct a sequence K_n' of compact subsets of Ω satisfying 1, 2 and 4 just as in the proof of Theorem A1.8. We then define $K_n = \hat{K}_n'$ and note that 1 continues to hold since the operation of forming the $A(\Omega)$-hull clearly preserves inclusions.

Notice that condition 4 implies that every compact subset of Ω is contained in K_n for n sufficiently large.

Choose an infinite subsequence (x_k) of (z_n) which possesses the property that $x_k \in K_{n(k)+1} \setminus K_{n(k)}$, $k \geq 1$, where the sequence $(n(k))$ is strictly increasing. We leave the construction of such a sequence as an easy exercise for the reader (note that $(K_{n+1} - K_n) \cap (z_n)$ is always finite). It is clearly enough to construct $f \in A(\Omega)$ which is unbounded on (x_k). For $k \geq 1$, let $L_k = K_{n(k)}$, $\|g\|_k = \|g\|_{L_k}$ and observe that the sequence (L_k) satisfies conditions 1 to 4 above.

We construct inductively a sequence $f_k \in A(\Omega)$ satisfying

$$f_k(x_k) = k+1+\sum_{j=1}^{k-1} |f_j(x_k)| \text{ and } \|f_k\|_k \leq 2^{-k+2} \qquad \ldots.(*)$$

Take $f_1 \equiv 2$. Suppose $f_1,\ldots,f_{k-1}$ are constructed. Since $x_k \notin L_k$, there exists $g \in A(\Omega)$ such that $|g(x_k)| > \|g\|_k$. Dividing by $g(x_k)$ we may assume

$$1 = g(x_k) > \|g\|_k \quad .$$

Choose $p(k)$ so large that

$$\|g^{p(k)}\|_k \leq 2^{-k+2}/(k+1+\sum_{j=1}^{k-1} |f_j(x_k)|) \; .$$

Set $f_k = (k+1+\sum_{j=1}^{k-1} |f_j(x_k)|)g^{p(k)}$. The inductive step is completed. We now define

$$f = \sum_{k=1}^{\infty} f_k \; .$$

Conditions (*) imply that the sum converges uniformly on L_k for $k \geq 1$ (compare with the proof of Theorem A1.8) and also that $|f(x_k)| \geq k-1$, $k \geq 1$. Since every compact subset of Ω is contained in L_k for k large enough it follows by Corollary 2.1.8 that $f \in A(\Omega)$. Since f is unbounded on (x_k) the proof is complete. □

Remarks.

1. The technique used to construct the function in the second part of the proof is very effective and we use it again shortly. Holomorphic convexivity is well adapted to the construction of analytic functions.

2. The sequence (K_n) of compact subsets of Ω constructed in the proof of Theorem 2.4.6 is called a *normal exhaustion* of Ω.

Example 11. Any open convex subset of $\mathbb{C}^n$ is a domain of holomorphy. Indeed, since $\hat{K} \subset K_c$, any open convex subset of $\mathbb{C}^n$ must be holomorphically convex. Of course, using the technique of the proof of Lemma 2.4.5, it can easily be proved directly that an open convex subset of $\mathbb{C}^n$ is a domain of holomorphy.

Remark. Holomorphic convexivity is much weaker than convexivity: Any open subset of $\mathbb{C}$ is holomorphically convex as are products of open subsets of $\mathbb{C}$.

Definition 2.4.7. Let $f \in A(\Omega)$. We say that Ω is the *domain of existence* of f, if f cannot be analytically continued across any point of the boundary of Ω.

Remarks.

1. Every domain of existence is trivially a domain of holomorphy.

2. If Ω is a domain in $\mathbb{C}$ then Ω is a domain of existence by Corollary 1.3.5.

Theorem 2.4.8. Every holomorphically convex domain Ω in $\mathbb{C}^n$ is the domain of existence of some analytic function on Ω.

Proof. Our proof is rather similar to that of Theorem 2.4.6.

Let (K_n) be a normal exhaustion of Ω and M be a discrete subset of Ω such that $\bar{M} \supset \partial\Omega$ (see the proof of Corollary 1.3.5 for the construction of M). For each $n \geq 1$, there exist only finitely many points of M in $K_{n+1} \setminus K_n$, say $x_1,\ldots,x_p$. As in the proof of Theorem 2.4.6 there exist $g_1,\ldots,g_p \in A(\Omega)$ such that

$$1 = g_1(x_1) = \ldots = g_p(x_p) > \|g_1\|_{K_n},\ldots,\|g_p\|_{K_n} .$$

By perturbing each g_j by a term $\theta(z - x_j)$, where $\theta \in \mathbb{C}^{n*}$ is small, it is clear that we can assume that $|g_j(x_k)| \neq 1$, $j \neq k$.

Set $a_{jk}(z) = (g_j(z) - g_j(x_k))/(1 - g_j(x_k))$, $j \neq k$. Taking sufficiently high powers of the g_j's, we may further require that

$$\|g_j\|_{K_n} \leq 2^{-n-p}/p \text{ and } \|a_{jk}\|_{K_n} \leq 2 .$$

We now define

$$f_n(z) = \sum_{j=1}^{p} \left(\prod_{k \neq j} a_{jk}(z) \right) g_j(z), \ z \in \Omega .$$

Our estimates imply that

$$f_n(x_i) = 1, \ 1 \leq i \leq p$$

$$\|f_n\|_{K_n} \leq 2^{-n} .$$

Define

$$f = \prod_{n=1}^{\infty} (1 - f_n)^n .$$

Since $\sum_{n=1}^{\infty} n2^{-n}$ is convergent, the infinite product converges uniformly on compact subsets of Ω and so $f \in A(\Omega)$. The convergence of the infinite product together with the fact that no f_n is identically equal to 1 implies that f is not identically zero.

Let $(x_k) \subset M$ be a sequence converging to $x \in \partial\Omega$. If $x_k \in K_{n(k)+1} \setminus K_{n(k)}$, f and its first n(k) derivatives vanish at x_k. Since $n(k) \to \infty$ as $k \to \infty$, we see that any analytic extension of f to a neighbourhood of x would have all its derivatives vanishing at x. But this implies, by uniqueness of analytic continuation, that $f \equiv 0$. Hence f does not extend across any point of $\partial\Omega$ and Ω is the domain of existence of f. □

Remark. Using the method of proof of Theorems 2.4.6, 2.4.8 it is not hard to show that if M is an infinite discrete subset of a holomorphically convex domain Ω, then there exists $f \in A(\Omega)$ which is unbounded on *every* infinite subset of M. Clearly this very strong form of Property (S) implies Theroem 2.4.8 as well as the non-trivial part of Theorem 2.4.6.

Let us summarise what we have proved so far. We have shown that for a domain Ω in $\mathbb{C}^n$, Property (S) is equivalent to holomorphic convexivity. We have also proved that if Ω is holomorphically convex then Ω is a domain of existence which in turn implies that Ω is a domain of holomorphy. We next show that these four properties are equivalent by proving that every domain of holomorphy is holomorphically convex.

First some notation. If $K \subset \Omega$ is compact, we let

$$\begin{aligned} d_\Omega(K) &= d(\partial\Omega, K) \\ &= \inf\{|z-\zeta| : z \in \partial\Omega,\ \zeta \in K\}. \end{aligned}$$

Provided that $\Omega \neq \mathbb{C}^n$, $d_\Omega(K) < \infty$. We remark that if $d(z,\partial\Omega) = r$ and $z \in \Omega$, then $D(z;r) \subset \Omega$ and indeed is the largest open polydisc $D(z;s)$ contained in Ω and centered at z.

Proposition 2.4.9. Let K be a compact subset of $\Omega \subset \mathbb{C}^n$ and $z_0 \in \hat{K}$. If $f \in A(\Omega)$, the power series for f at z_0,

$$f(z) = \sum_m \frac{1}{m!} \partial^m f(z_0)(z - z_0)^m ,$$

converges on $D(z_0 : d_\Omega(K))$.

Proof. Choose r, $0 < r < d_\Omega(K)$ and define $K_r = \bigcup_{z \in K} \bar{D}(z;r)$. K_r is certainly compact and contined in Ω. Let $M = \|f\|_{K_r}$. By Corollary 2.1.8 applied to $\bar{D}(z;r)$ we have

$$|\partial^m f(z)| \leq Mm!r^{-|m|}$$

for $z \in K$ ($|m| = m_1 + \ldots + m_m$). Hence

$$\|\partial^m f\|_K \leq Mm!r^{-|m|}.$$

Since $\partial^m f \in A(\Omega)$ we have, by definition of $\hat{K}$,

$$\|\partial^m f\|_{\hat{K}} \leq Mm!r^{-|m|}.$$

Since $z_0 \in \hat{K}$, this estimate implies

$$|\partial^m f(z_0)| \le Mm! r^{-|m|}$$

and so the series $\sum \frac{1}{m!} \partial^m f(z_0)(z - z_0)^m$ converges for $|z - z_0| < r$. Since this is true for all $r < d_\Omega(K)$, the Proposition follows. □

Theorem 2.4.10. (Cartan-Thullen). A domain of holomorphy is holomorphically convex.

Proof. It is clearly enough to prove that for any compact subset K of a domain of holomorphy Ω, $d_\Omega(K) = d_\Omega(\hat{K})$. Obviously $d_\Omega(K) \ge d_\Omega(\hat{K})$. Suppose that for some compact subset K of Ω, $d_\Omega(K) > d_\Omega(\hat{K})$. Pick $z_0 \in \hat{K}$ such that $d(z_0, \partial\Omega) < d_\Omega(K)$. By Proposition 2.4.9, $\sum \frac{1}{m!} \partial^m f(z_0)(z - z_0)^m$ converges on $D(z_0; d_\Omega(K))$. But $D(z_0; d_\Omega(K)) \not\subset \Omega$ (see remarks immediately preceding Proposition 2.4.9). Hence we have constructed an analytic extension of every $f \in A(\Omega)$ to some open set not contained in Ω. Contradiction. Hence Ω is holomorphically convex. □

Examples.

12. Suppose $\Omega, \Omega' \subset \mathbb{C}^n$ and $f: \Omega \to \Omega'$ is biholomorphic. If one of Ω, Ω' is a domain of holomorphy so is the other. This follows since holomorphic convexivity is obviously preserved by biholomorphic maps. Notice that it is *not* obvious from the definitions that domains of holomorphy or existence are preserved by biholomorphic maps.

13. Suppose that $\Omega \subset \mathbb{C}^n$, $\Omega' \subset \mathbb{C}^m$ are domains of holomorphy and $f: \Omega \to \mathbb{C}^m$ is analytic. Then $f^{-1}(\Omega')$ is a domain of holomorphy. We show that $f^{-1}(\Omega')$ possesses property (S). Let (z_n) be a sequence of points of $f^{-1}(\Omega')$ converging to some point $z \in \partial f^{-1}(\Omega')$. If $z \in \partial\Omega$, then there exists $g \in A(\Omega)$ which is unbounded on (z_n). Restricting to $f^{-1}(\Omega')$, we have found an analytic function on $f^{-1}(\Omega')$ unbounded on (z_n). If $z \notin \partial\Omega$, then $(f(z_n))$ converges to $f(z) \in \mathbb{C}^m$. Clearly, $f(z) \in \partial\Omega'$, and so there exists $g \in A(\Omega')$ which is unbounded on $(f(z_n))$. But then $gf \in A(f^{-1}(\Omega'))$ is unbounded on (z_n). We remark that we can weaken the conditions on Ω to require only that Ω possesses property (S) at all points of $\partial\Omega \cap \partial f^{-1}(\Omega')$.

14. Let $\Omega \subset \mathbb{C}^n$ be a domain of holomorphy and $f_1, \ldots, f_q \in A(\Omega)$. Then $P = \{z \in \Omega: |f_j(z)| < 1,\ j = 1, \ldots, q\}$ is a domain of holomorphy. Indeed,

set $f = (f_1,\dots,f_q): \Omega \to \mathbb{C}^q$. Then $P = f^{-1}(D(0;1))$ and example 13 applies. We recall that P is an analytic polyhedron.

15. Let $\Omega_i \subset \mathbb{C}^n$, $i \in I$, be domains of holomorphy. Then $\Omega = \text{Interior}\,(\bigcap_{i\in I} \Omega_i)$ is a domain of holomorphy. We prove that Ω is holomorphically convex. Let K be a compact subset of Ω and $\hat{K}$ and $\hat{K}_i$ respectively denote the $A(\Omega)$- and $A(\Omega_i)$-hulls of K. Certainly $\hat{K} \subset \hat{K}_i$, $i \in I$. Hence, with the notation of Theorem 2.4.10, $d_\Omega(K) \le d_{\Omega_i}(\hat{K})$, $i \in I$. This being true for all $i \in I$, we must have $d_\Omega(\hat{K}) \ge d_\Omega(K)$ and so K is a compact subset of Ω.

Exercises.

1. Let Ω be a domain of holomorphy and M be an infinite discrete subset of Ω. Show that there exists an analytic function on Ω which is unbounded on every infinite subset of M (see the remarks following Theorem 2.4.8).

2. Let Ω be holomorphically convex. Show that there exist normal exhaustions (K_n) of Ω that satisfy the additional property $K_n \subset \overset{\circ}{K}_{n+1}$, $n \ge 1$.

§5. Pseudoconvexivity

The results of §4 show that there is a close parallel between the convexivity of domains in $\mathbb{R}^n$ and holomorphic convexivity of domains in $\mathbb{C}^n$. In this section we shall explore this analogy further. First we shall review some facts about convex domains in $\mathbb{R}^n$. A standard reference is Bonnesen and Fenchel [1].

Let Ω be a proper subdomain of $\mathbb{R}^n$ and suppose $\partial\Omega$ is a C^r submanifold of $\mathbb{R}^n$, $r \ge 1$. Using partitions of unity it is not hard to construct $\phi \in C^r(\mathbb{R}^n,\mathbb{R})$ satisfying

A) $\Omega = \{x \in \mathbb{R}^n : \phi(x) < 0\}$

B) $\partial\Omega = \{x \in \mathbb{R}^n : \phi(x) = 0\}$

C) $d\phi \neq 0$ on $\partial\Omega$.

We call a map ϕ satisfying these conditions a (C^r-) defining function for Ω. The following technical lemma will be important in defining boundary invariants of Ω.

Lemma 2.5.1. Let Ω be a proper subdomain of $\mathbb{R}^n$ with C^2 boundary and suppose that ϕ and θ are defining functions for Ω. Then there exists $h \in C^1(\mathbb{R}^n, \mathbb{R})$ satisfying

a) $\phi = h.\theta$

b) h is strictly positive

c) h is C^2 off $\partial\Omega$

d) Given $1 \leq i, j \leq n$, let

$$\begin{aligned} a_{ij}(x) &= d(x,\partial\Omega)\,\frac{\partial^2 h}{\partial x_i \partial x_j}(x),\ x \notin \partial\Omega \\ &= 0,\ x \in \partial\Omega. \end{aligned}$$

Then a_{ij} is continuous.

($d(x,\partial\Omega)$ is the distance from x to $\partial\Omega$, relative to some norm on $\mathbb{R}^n$).

Proof. Clearly we may define $h = \phi/\theta$ on $\mathbb{R}^n\backslash\partial\Omega$ and h is C^2 on $\mathbb{R}^n\backslash\partial\Omega$. The problem is to show that h extends as a positive C^1 function across $\partial\Omega$ satisfying property d). Fix $z \in \partial\Omega$. It is enough to find an open neighbourhood U of z in $\mathbb{R}^n$ and $h \in C^1(U,\mathbb{R})$ such that h satisfies properties a) - d) in U. Choose local coordinates on some open neighbourhood U of z so that z corresponds to zero, $\partial\Omega \cap U = \{x \in U: x_1 = 0\}$ and, in the new coordinates, U is convex. In the new coordinate system we have $\phi(0,x_2,\ldots,x_n) = 0$ and $\frac{\partial\phi}{\partial x_1}(0,x_2,\ldots,x_n) \neq 0$ for $(0,x_2,\ldots,x_n) \in U$. Similarly for θ. Now

$$\begin{aligned} \phi(x_1,\ldots,x_n) &= \int_0^1 \frac{d}{dt}(\phi(tx_1,\ldots,x_n))dt \\ &= \int_0^1 x_1 \frac{\partial\phi}{\partial x_1}(tx_1,\ldots,x_n)dt \\ &= x_1\tilde{\phi}(x_1,\ldots,x_n). \end{aligned}$$

Since $\frac{\partial\phi}{\partial x_1}(0,x_2,\ldots,x_n) \neq 0$, $\tilde{\phi}(0,x_2,\ldots,x_n) \neq 0$ on $\partial\Omega \cap U$. Furthermore $\tilde{\phi}$ is C^1 on U. Similarly, $\theta(x_1,\ldots,x_n) = x_1\tilde{\theta}(x_1,\ldots,x_n)$ and so we may extend h to a C^1 function on U by setting $h = \tilde{\phi}/\tilde{\theta}$.

For property d), it is enough to show that the function

$$A_{ij}(x) = x_1 \frac{\partial^2 \tilde{\phi}}{\partial x_i \partial x_j}(x), \; x \notin \partial\Omega \cap U$$
$$= 0, \; x \in \partial\Omega \cap U$$

is continuous on U for $1 \le i, j \le n$. We shall prove this is so in case $i = j = 1$, leaving the remaining (easier) cases $i = 1$, $j \ne 1$ and $i,j \ne 1$ to the reader.

Since $\tilde{\phi}$ is C^2 provided $x_1 \ne 0$, we may differentiate the identity $\phi = x_1\tilde{\phi}$ twice to obtain

$$\frac{\partial\phi}{\partial x_1} = \tilde{\phi} + x_1 \frac{\partial\tilde{\phi}}{\partial x_1}$$

$$\frac{\partial^2\phi}{\partial x_1^2} = 2\frac{\partial\tilde{\phi}}{\partial x_1} + x_1 \frac{\partial^2\tilde{\phi}}{\partial x_1^2}, \; x_1 \ne 0 .$$

Eliminating $\partial\tilde{\phi}/\partial x_1$ from the second relation we obtain

$$\text{*)....} \qquad x_1\tilde{\phi}_{11} = \frac{\frac{1}{2}x_1^2\phi_{11} - x_1\phi_1 + \phi}{\frac{1}{2}x_1^2}, \; x_1 \ne 0 ,$$

where we have used the abbreviated notation ϕ_{11} and ϕ_1 for $\partial^2\phi/\partial x_1^2$ and $\partial\phi/\partial x_1$ respectively.

By Taylor's theorem, there exists ξ, $0 < \xi < 1$, such that

$$\phi(x_1,x_2,\ldots,x_n) = \phi(0,x_2,\ldots,x_n) + x_1\phi_1(0,x_2,\ldots,x_n) + \tfrac{1}{2}x_1^2\phi_{11}(\xi x_1,x_2,\ldots,$$

$$= x_1\phi_1(0,x_2,\ldots,x_n) + \tfrac{1}{2}x_1^2\phi_{11}(\xi x_1,x_2,\ldots,x_n).$$

Substituting, we see that the numerator of the right hand side of *) is equal to

$$\tfrac{1}{2}x_1^2\phi_{11}(x_1,\ldots,x_n) - x_1\phi_1(x_1,\ldots,x_n) + x_1\phi_1(0,\ldots,x_n) + \tfrac{1}{2}x_1^2\phi_{11}(\xi x_1,\ldots,x_n) .$$

By the Mean value theorem there exists ρ, $0 < \rho < 1$, such that

$$x_1\phi_1(x_1,\ldots,x_n) - x_1\phi_1(0,\ldots,x_n) = x_1^2\phi_{11}(\rho x_1,\ldots,x_n) .$$

Substituting in the above expression for the numerator of the right hand side of *), we arrive at the following formula for $x_1\tilde{\phi}_{11}$:

$$x_1\tilde{\phi}_{11}(x_1,\ldots,x_n) = \frac{x_1^2}{2}(\phi_{11}(x_1,\ldots,x_n)+\phi_{11}(\xi x_1,\ldots,x_n)-2\phi_{11}(\rho x_1,\ldots,x_n)),$$

provided $x_1 \neq 0$. The continuity of A_{11} now follows from the continuity of ϕ_{11}. □

Remark. It must be emphasised that even though ϕ and θ are C^2, h may not be C^2. For example, if $n = 1$, $\phi(x) = x+x^2|x|$, $\theta(x) = x$, then ϕ/θ will only be of class C^1.

Let $Q_n(\mathbb{R})$ denote the space of real quadratic forms on $\mathbb{R}^n$. Taking the standard basis on $\mathbb{R}^n$, we may identify $Q_n(\mathbb{R})$ with the space of $n \times n$ real symmetric matrices and we shall give $Q_n(\mathbb{R})$ the topology induced from that on the space of $n \times n$ real matrices. If $q \in Q_n(\mathbb{R})$ corresponds to $[a_{ij}]$, we set

$$q(v) = \sum_{i,j=1}^{n} a_{ij}v_iv_j, \quad v = (v_1,\ldots,v_n) \in \mathbb{R}^n.$$

Let $\phi \in C^2(\mathbb{R}^n,\mathbb{R})$. The *Hessian* of ϕ is the continuous map $H(\phi)\colon \mathbb{R}^n \to Q_n(\mathbb{R})$ defined by

$$H(\phi)(x) = \left[\frac{\partial^2\phi}{\partial x_i \partial x_j}(x)\right].$$

Avoiding coordinates, we may equivalently define $H(\phi)(x) = D^2\phi_x$ - the second derivative of ϕ at x.

Suppose $\Omega \subset \mathbb{R}^n$ has C^2 boundary and that ϕ is a C^2 defining function for Ω. We say that $H(\phi)$ is positive semi-definite on tangent vectors to $\partial\Omega$ if for all $x \in \partial\Omega$

$$H(\phi)(x)(v) \geq 0, \text{ for all } v \in \mathbb{R}^n \text{ such that } d\phi(x)(v) = 0 .$$

We say $H(\phi)$ is positive definite on tangent vectors to $\partial\Omega$ if for all $x \in \partial\Omega$

$$H(\phi)(x)(v) > 0, \text{ for all non-zero } v \in \mathbb{R}^n \text{ such that } d\phi(x)(v) = 0 .$$

Lemma 2.5.2. Suppose Ω is a proper subdomain of $\mathbb{R}^n$ with C^2 boundary. If there exists a C^2 defining function ϕ for Ω such that $H(\phi)$ is positive (semi-) definite on tangent vectors to $\partial\Omega$ then the same is true for the Hessian of every C^2 defining function for Ω.

Proof. Suppose θ is another C^2 defining function for Ω. By Lemma 2.5.1 there exists a strictly positive C^1 function h on $\mathbb{R}^n$ such that $\phi = h.\theta$. Since h is C^2 off $\partial\Omega$, we certainly have

$$H(\phi)(x) = \theta(x)H(h)(x) + 2(d\theta(x), dh(x)) + h(x)H(\theta)(x), \ x \in \mathbb{R}^n \backslash \partial\Omega .$$

By part d) of Lemma 2.5.1, the function $x \longmapsto \theta(x)H(h)(x)$ extends to a continuous function on $\mathbb{R}^n$ which vanishes on $\partial\Omega$. Hence

$$H(\phi)(x)(v) = 2(dh(x)(v)d\theta(x)(v)) + h(x)H(\theta)(x)(v) ,$$

$x \in \partial\Omega$, $v \in \mathbb{R}^n$. If v is a tangent vector to $\partial\Omega$, $d\theta(x)(v) = 0$ and so we see that

$$H(\phi)(x)(v) = h(x)H(\theta)(x)(v), \ x \in \partial\Omega, \ v \text{ a tangent vector to } \partial\Omega \text{ at } x.$$

Since $h(x) > 0$, the result follows. □

Theorem 2.5.3. Let Ω be a proper subdomain of $\mathbb{R}^n$ with C^2 boundary. Then Ω is convex if and only if for some defining function ϕ for Ω, $H(\phi)$ is positive semi-definite on tangent vectors to $\partial\Omega$.

Proof. We shall prove that the positive semi-definiteness of $H(\phi)$ on tangent vectors to $\partial\Omega$ implies the convexivity of Ω and leave the converse as an easy exercise for the reader. We first remark that it is sufficient to prove that $\bar{\Omega}$ is locally convex. That is, for each $z \in \partial\Omega$ there exists a (convex) neighbourhood N of z in $\mathbb{R}^n$ such that $N \cap \bar{\Omega}$ is convex. A proof of this well known characterisation of convex domains may be found in Valentine [1]. Take the Euclidean norm on $\mathbb{R}^n$ and let $\bar{E}_r(x)$ denote the closed disc, radius r and centre x in $\mathbb{R}^n$. For local convexivity it is sufficient to show that for each $z \in \partial\Omega$, there exists $r > 0$ such that $T_y\partial\Omega \cap \bar{E}_r(z) \cap \Omega = \emptyset$, for all $y \in \partial\Omega \cap \bar{E}_r(z)$. Indeed, this condition implies that $\bar{E}_r(z) \cap \bar{\Omega}$ is convex (see the characterisation of closed convex hull given in §4).

Fix $z \in \partial\Omega$. By an affine linear change of coordinates we may assume that $z = 0$, the tangent plane to $\partial\Omega$ at 0 is the hyperplane $x_n = 0$ and $\frac{\partial\phi}{\partial x_n}(0) < 0$.

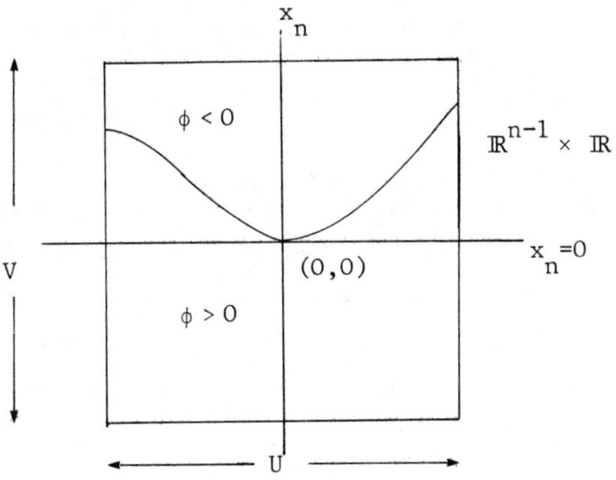

Figure 5.

By the implicit function theorem we can choose an open neighbourhood $U \times V$ of $(0,0) \in \mathbb{R}^{n-1} \times \mathbb{R}$ such that $(U \times V) \cap \partial\Omega$ can be represented as the graph of a C^2 function $\psi: U \to V$. That is, $\phi(x,\psi(x)) = 0$, $x \in U$, and $(U \times V) \cap \partial\Omega = \{(x,\psi(x)): x \in U\}$.

Differentiating the identity $\phi(x,\psi(x)) = 0$ twice we obtain

$$0 = D_1^2\phi_X + 2D_{12}\phi_X.D\psi_x + D_2^2\phi_X.(D\psi_x,D\psi_x) + D_2\phi_X.D^2\psi_x, \quad x \in U .$$

Here we have set $X = (x,\psi(x))$ and we follow the derivative notation of Field [1].

Now any tangent vector to graph(ψ) at x is of the form $(v,D\psi_x(v))$, $v \in \mathbb{R}^{n-1}$. Set $V = (v,D\psi_x(v))$, $v \in \mathbb{R}^{n-1}$. Evaluating the above quadratic form on V, we obtain

$$0 = H(\phi)(X)(V) + \frac{\partial\phi}{\partial x_n}(X)D^2\psi_x(v,v), \quad v \in \mathbb{R}^{n-1}, \ x \in U .$$

Now $H(\phi)(X)(V) \geq 0$ by assumption. Furthermore, since $\frac{\partial\phi}{\partial x_n}(0) < 0$, we can find $s > 0$ such that $\frac{\partial\phi}{\partial x_n}(X) < 0$, $\|(x_1,\ldots,x_{n-1})\| < s$ ($\|\ \|$ denotes the Euclidean norm). Therefore, $D^2\psi_x(v,v) \geq 0$, $\|x\| < s$, $v \in \mathbb{R}^{n-1}$.

Pick a unit vector $u \in \mathbb{R}^{n-1}$ and let L_u denote the line through the origin of $\mathbb{R}^{n-1}$ defined by u. Let $\psi^u = \psi|L_u$. The positivity of $D^2\psi_x$ implies that

$$\frac{d^2\psi^u}{du^2}(x) \geq 0, \; x \in L^u, \; \|x\| < s.$$

As is well known this implies that the graph of ψ^u is convex. In particular, $L^u \cap \text{graph}(\psi^u) \neq \emptyset$. This is true for all unit vectors u in $\mathbb{R}^{n-1}$ and so the open disc in $\mathbb{R}^{n-1}$ of radius s, centre 0, does not intersect $\Omega \cap \bar{E}_s(0)$. A straightforward argument based on the continuity of grad(ϕ) shows that we can extend this result to find r, $0 < r \leq s$, such that $T_y\partial\Omega \cap \bar{E}_r(0) \cap \Omega = \emptyset$, for all $y \in \partial\Omega \cap \bar{E}_r(0)$. As we pointed out above, this is sufficient to prove the convexivity of Ω. □

Definition 2.5.4. Let Ω be a proper subdomain of $\mathbb{R}^n$ with C^2 boundary. We say that Ω is *strictly convex* if for some defining function ϕ for Ω, $H(\phi)$ is positive definite on tangent vectors to $\partial\Omega$.

Proposition 2.5.5. Let Ω be a strictly convex bounded domain in $\mathbb{R}^n$ with C^2 boundary. Then there exists a defining function ϕ for Ω such that $H(\phi)(x)$ is a positive definite quadratic form for all $x \in \partial\Omega$.

Proof. Let θ be any C^2 defining function for Ω. For $\lambda \in \mathbb{R}$, set $\phi_\lambda = \theta e^{\lambda\theta}$. Clearly for all values of λ, ϕ_λ is a C^2 defining function on Ω which is positive definite on tangent vectors to $\partial\Omega$. Computing the Hessian of ϕ_λ and restricting to $\partial\Omega$ we find

$$H(\phi_\lambda)(x)(v) \;=\; H(\theta)(x)(v) + 2\lambda(d\theta(x)(v))^2, \; v \in \mathbb{R}^n, \; x \in \partial\Omega .$$

Restrict $H(\phi_\lambda)$ to the compact set $\Gamma = \partial\Omega \times S^{n-1}$, where S^{n-1} denotes the unit sphere in $\mathbb{R}^n$. Let $K = \{(x,u) \in \Gamma, \, d\theta(x)(v) = 0\}$(unit tangent bundle of $\partial\Omega$). $H(\theta)$ is strictly positive on K and so, by continuity, is strictly positive on some open neighbourhood U of K in Γ. Let m be the infimum of $2(d\theta(x)(v))^2$ over $\Gamma \setminus U$ and M the supremum of $H(\theta)(x)(v)$ over $\Gamma \setminus U$. Since the compact set $\Gamma \setminus U$ does not meet the zero set of $d\theta$, $m > 0$. Taking $\lambda_0 = 2M/m$, we see immediately that $H(\phi_\lambda)$ is strictly positive on Γ, $\lambda \geq \lambda_0$. □

This completes our review of convex domains in $\mathbb{R}^n$. Our aim is now to obtain a characterisation of domains of holomorphy with C^2 boundary in terms of local properties of the boundary. For the remainder of this section Ω will always denote an open, connected subset of $\mathbb{C}^n$.

We let $H_n(\mathbb{C})$ denote the space of Hermitian quadratic forms on $\mathbb{C}^n$. Taking the standard complex basis on $\mathbb{C}^n$, we may identify $H_n(\mathbb{C})$ with the space of $n \times n$ Hermitian matrices and we give $H_n(\mathbb{C})$ the corresponding topology.

Definition 2.5.6. Let D be an open subset of $\mathbb{C}^n$, and $\phi \in C^2(D,\mathbb{R})$. The *Levi form* of ϕ is the map

$$L(\phi): D \to H_n(\mathbb{C})$$

defined by

$$L(\phi)(z) = \left[\frac{\partial^2\phi}{\partial z_i \partial \bar{z}_j}\right], \ z \in D .$$

Remarks.

1. Given $Z = (Z_1,\ldots,Z_n) \in \mathbb{C}^n$, $z \in D$,

$$L(\phi)(z)(Z) = \sum_{i,j=1}^{n} \frac{\partial^2\phi}{\partial z_i \partial \bar{z}_j}(z)\, Z_i \bar{Z}_j .$$

2. In Chapter 5 we shall give an invariant definition of the Levi form.

As we shall soon see the Levi form is the complex analogue of the Hessian.

Lemma 2.5.7. Let Ω, Ω' be domains in $\mathbb{C}^n$ and $h: \Omega \to \Omega'$ be biholomorphic. If $\phi \in C^2(\Omega',\mathbb{R})$ then

$$L(\phi h) = L(\phi)(Dh,\overline{Dh}).$$

(Here $[Dh] = [\partial h_i/\partial z_j]$; $[\overline{Dh}] = [\overline{\partial h_i/\partial z_j}]$).

Proof. We have to show that for $1 \le i, j \le n$,

$$\frac{\partial^2 \phi h}{\partial z_i \partial \bar{z}_j} = \sum_{\alpha,\beta=1} \frac{\partial^2\phi}{\partial z_\alpha \partial \bar{z}_\beta} \frac{\partial h_\alpha}{\partial z_i} \overline{\frac{\partial h_\beta}{\partial z_j}} .$$

This is a straightforward computation using Proposition 2.1.1 and we omit details. □

Lemma 2.5.7 shows that the Levi form is invariant under holomorphic changes of variables. Of course, the Hessian is only invariant under affine linear changes of variables.

Lemma 2.5.8. Let $\phi \in C^2(\Omega,\mathbb{R})$. Then

$$L(\phi) = \tfrac{1}{4}(H(\phi) + H(\phi)(J)) .$$

That is, for $z \in \Omega$, $Z \in \mathbb{C}^n$,

$$L(\phi)(z)(Z) = \tfrac{1}{4}(H(\phi)(z)(Z) + H(\phi)(z)(JZ)) ,$$

where on the right hand side we use the standard identification of $\mathbb{C}^n$ with $\mathbb{R}^{2n}$ and regard $Z \in \mathbb{R}^{2n}$ (J denotes the standard complex structure on $\mathbb{C}^n$ (see §5, Chapter 1)).

Proof. A straightforward, if tedious, calculation and we omit details. □

Example 1. Let $\Omega \subset \mathbb{C}^n$ be a bounded convex domain. We know from §4 that every biholomorphic image of Ω is a domain of holomorphy. Suppose that Ω has C^2 boundary and that $g \in C^2(\bar{\Omega},\mathbb{C}^n)$ is a diffeomorphism of some neighbourhood U of $\bar{\Omega}$ into $\mathbb{C}^n$ which maps Ω biholomorphically onto a domain $\Omega' \subset \mathbb{C}^n$. Certainly Ω' has C^2 boundary and $\partial\Omega' = g(\partial\Omega)$. Set $h = (g|U)^{-1}$. If ϕ is a defining function for Ω, then ϕh is a defining function for Ω' (ϕh is only defined on a neighbourhood of Ω', but we can always extend to a C^2 function on the whole of $\mathbb{C}^n$ which is strictly positive outside Ω'). Since Ω is convex, $H(\phi)$ is positive semi-definite on tangent vectors to $\partial\Omega$. Suppose Z is a holomorphic tangent vector to Ω' at x. That is,

$$\sum_{i=1}^{n} \frac{\partial(\phi h)}{\partial z_i}(x)Z_i = 0, \quad Z = (Z_1,\ldots,Z_n) .$$

Taking real and imaginary parts, we find that Z and JZ are *real* tangent vectors to $\partial\Omega'$ at x. Therefore, by Lemmas 2.5.7, 2.5.8,

$$L(\phi h)(x)(Z) = \tfrac{1}{4}(H(\phi)(x)(Dh_x(Z)) + H(\phi)(x)(JDh_x(Z)) \geq 0 .$$

In other words, the Levi form is positive semi-definite on holomorphic tangent vectors to $\partial\Omega'$. Furthermore, if Ω is strictly convex, the Levi form will be positive definite on holomorphic tangent vectors to $\partial\Omega'$.

Remark. Of course a domain of holomorphy need not be the biholomorphic image of a convex set. However, if Ω is a domain in $\mathbb{C}^n$ with C^2 boundary and defining function ϕ such that $L(\phi)$ is positive definite at a point $z \in \partial\Omega$, then we may clearly make a local ($\mathbb{C}$-linear) holomorphic change of coordinates on some neighbourhood U of z such that, in the new coordinates, $U \cap \Omega$ is convex. In particular, if $L(\phi)$ is positive definite at all points of $\partial\Omega$, we see that Ω is locally the biholomorphic image of a convex set. Consequently, by example 1, Ω will locally be a domain of holomorphy. That is, for every $z \in \bar{\Omega}$, there will exist an open neighbourhood U of z in $\mathbb{C}^n$ such that $U \cap \Omega$ is a domain of holomorphy. However, if $L(\phi)$ is not positive definite and only positive semi-definite, even at just one point of $\partial\Omega$, it is not generally the case that Ω is locally the biholomorphic image of a convex set (it will however, be locally a domain of holomorphy). For examples and further references we refer the reader to the survey article by Y-T Siu [1], especially §7. We should mention that the question of boundary invariants of domains with real analytic boundary is currently a topic of much interest. See, for example, the paper of Chern and Moser [1].

Definition 2.5.9. Let Ω be a domain in $\mathbb{C}^n$ with C^2 boundary. Suppose that there exists a defining function ϕ for Ω such that $L(\phi)$ is positive semi-definite on holomorphic tangent vectors to $\partial\Omega$. Then we say that Ω is *Levi pseudoconvex* (abbreviated L.p.). In case $L(\phi)$ is positive definite on holomorphic tangent vectors to $\partial\Omega$, we say Ω is *strictly Levi pseudoconvex* (abbreviated s.L.p.).

Exactly as we did for convex subsets of $\mathbb{R}^n$ we may prove

Lemma 2.5.10. Suppose Ω is a domain in $\mathbb{C}^n$ with C^2 boundary. If Ω is L.p. then $L(\phi)$ is positive semi-definite on holomorphic tangent vectors to $\partial\Omega$ for *all* C^2 defining functions ϕ for Ω. Similarly for s.L.p. domains.

Lemma 2.5.11. If $\Omega \subset \mathbb{C}^n$ is a bounded, strictly Levi pseudoconvex domain with C^2 boundary then there exists a defining function ϕ for such that $L(\phi)$ is positive definite on $\partial\Omega$.

Just as for convexity, Levi pseudoconvexity is a local property of the boundary. More precisely we have

Proposition 2.5.12. Let Ω be a bounded domain in $\mathbb{C}^n$ with C^2 boundary. Suppose that for each $x \in \partial\Omega$, there exists an open neighbourhood U of x and $\phi \in C^2(U,\mathbb{R})$ such that

a) $\Omega \cap U = \{x \in U: \phi(x) < 0\}$.

b) $d\phi \neq 0$ on $\partial\Omega \cap U$.

c) $L(\phi)(y)(v) \geq 0$ for all $y \in \partial\Omega \cap U$ and holomorphic tangent vectors v to $\partial\Omega$ at y.

Then Ω is Levi pseudoconvex. If in c), $L(\phi)$ is positive definite on holomorphic tangent vectors then Ω is strictly Levi pseudoconvex.

Proof. Since $\partial\Omega$ is compact, we can find finitely many open subsets U_j of $\mathbb{R}^n$ and $\phi_j \in C^2(U_j,\mathbb{R})$ such that $\partial\Omega \subset \cup U_j$ and the ϕ_j satisfy the conditions of the proposition. Set $U = \cup U_j$ and choose $\theta_i \in C_c^\infty(U_i)$ such that $\theta_i \geq 0$ and $\bigcup_i \text{Interior}(\text{supp}(\theta_i)) \supset \partial\Omega$. Define $\tilde{\phi} = \Sigma\theta_i\phi_i \in C_c^\infty(U,\mathbb{R})$. Since $\tilde{\phi}(x) < 0$ if $x \in \Omega \cap U$ and $\tilde{\phi}(x) = 0$ if $x \in \partial\Omega$, we may choose $\psi \in C^\infty(\mathbb{C}^n,\mathbb{R})$ such that $\psi \equiv 0$ on a neighbourhood of $\partial\Omega$ and $\Omega = \{x \in \mathbb{C}^n: (\tilde{\phi}+\psi)(x) < 0\}$, $\partial\Omega = \{x \in \mathbb{C}^n: (\tilde{\phi}+\psi)(x) = 0\}$. Set $\phi = \tilde{\phi}+\psi$. On $\partial\Omega$,

$$\begin{aligned} d\phi &= d\tilde{\phi} = \Sigma\theta_i d\phi_i \\ &\neq 0 \quad . \end{aligned}$$

Computing $L(\phi)$, we find that on $\partial\Omega$

$$\begin{aligned} L(\phi)(x)(v) &= \textstyle\sum\theta_i(x)L(\phi_i)(x)(v), \text{ if } v \text{ is a holomorphic tangent vector to } \partial\Omega \text{ at } x, \\ &\geq 0 \, . \end{aligned}$$

□

Theorem 2.5.13 (E.E. Levi). Suppose Ω is a holomorphically convex domain in $\mathbb{C}^n$ with C^2 boundary. Then Ω is Levi pseudoconvex.

Proof. Let ϕ be a defining function for Ω and suppose that at some point $x \in \partial\Omega$, $L(\phi)$ is not positive semi-definite on holomorphic tangent vectors to $\partial\Omega$ at x. By a complex affine linear change of coordinat we may suppose that $x = 0$, $d\phi(0) = dx_1(0)$ and $L(\phi)(u) < 0$, where u is the unit basis vector in the z_2-direction. Thus, if for $1 \le i, j \le n$ we set $\phi_1(0) = \frac{\partial\phi}{\partial z_i}(0)$, $\phi_{\bar{j}}(0) = \frac{\partial\phi}{\partial \bar{z}_j}(0)$, $\phi_{ij}(0) = \frac{\partial^2\phi}{\partial z_i \partial z_j}(0)$, etc., we will have

$$\phi_{\bar{1}}(0) = \phi_1(0) = \tfrac{1}{2};\ \phi_i(0) = \phi_{\bar{i}}(0) = 0,\ i > 1:\ \phi_{2\bar{2}}(0) < 0 .$$

For $t \in \mathbb{R}$, we define the analytic embedding $\phi_t : \mathbb{C} \to \mathbb{C}^n$ by

$$\phi_t(s) = (-2\phi_{22}(0)s^2 - t, s, 0, 0, \ldots, 0) .$$

For $r > 0$, set $V_t(r) = \phi_t(D_r(0))$. $V_t(r)$ is an embedded disc in $\mathbb{C}^n$. We observe that $0 \in V_0(r)$. We claim that there exists $r > 0$ such that

$$\overline{V_t(r)} \subset \Omega,\ t \in (0,r]$$

$$\overline{V_0(r)} \setminus \{0\} \subset \Omega .$$

On the assumption that such an r exists we now show that Ω cannot be holomorphically convex. Define

$$K = \bigcup_{0 \le t \le r} \phi_t(\partial D_r(0)) = \bigcup_{0 \le t \le r} \partial V_t(r)$$

K is a compact subset of Ω. If $f \in A(\Omega)$, $0 < t \le r$, then $f\phi_t \in A(D_r(0))$ and is continuous on $\overline{D_r(0)}$. Hence, by the maximum principle, $\|f\phi_t\|_{\overline{D_r(0)}} = \|f\phi_t\|_{\partial D_r(0)}$. This implies that $\overline{V_t(r)} \subset \hat{K}$, $0 < t \le r$. By continuity, $\overline{V_0(r)} \setminus \{0\} \subset \hat{K}$. But therefore $\hat{K}$ cannot be compact and so Ω is not holomorphically convex.

To construct r we proceed as follows. By Taylor's theorem we may write

$$\phi(z) = x_1 + \sum_{i,j} (\phi_{ij}(0)z_i z_j + \phi_{i\bar{j}}(0)z_i\bar{z}_j + \phi_{\bar{i}\bar{j}}(0)\bar{z}_i\bar{z}_j) + o(|z|^2) .$$

Restricting ϕ to the sets V_t, we may regard ϕ as a function of s and t. Substituting we find

$$\phi(s,t) = \phi_{2\bar{2}}(0)|s|^2 + o(|s|^2) + tg(t,s) ,$$

where g is a continuous function of s and t satisfying $g(0,0) = -1$. Since $\phi_{2\bar{2}}(0) < 0$, we see that we can choose $r > 0$ so that $\phi(s,t) < 0$ for $t \in (0,r]$, $|s| \le r$ or $t = 0$, $0 < |s| \le r$. □

We may now state one of the most famous problems in several complex variables.

Levi's problem (E.E. Levi [1]). Show that every Levi pseudoconvex domain in $\mathbb{C}^n$ is a domain of holomorphy.

Levi's problem was answered affirmatively by Oka [1] in case $n = 2$ and later, independently, by Bremermann [1], Norguet [1] and Oka [2] for general n. We shall give a proof of a generalisation of Levi's problem, due to Grauert, in Chapter 7.

For the remainder of this section we shall investigate some alternative pseudoconvexivity definitions. First we need a technical lemma, the proof of which was shown to us by D.B.A. Epstein.

Lemma 2.5.14. Let Ω be a bounded domain in $\mathbb{R}^n$ with C^r boundary, $r \ge 2$. Define $d: \mathbb{R}^n \to \mathbb{R}$ by

$$\begin{aligned} d(x) &= d(x,\partial\Omega),\ x \in \Omega \\ &= -d(x,\partial\Omega),\ x \notin \Omega, \end{aligned}$$

where distances are computed relative to the Euclidean metric on $\mathbb{R}^n$. Then d is C^r on some neighbourhood of $\partial\Omega$ in $\mathbb{R}^n$.

Proof. Let $n(z)$ denote the unit inward normal to $\partial\Omega$ at $z \in \partial\Omega$ and $\phi: \partial\Omega \times \mathbb{R} \to \mathbb{R}^n$ be the map $\phi(z,\lambda) = z + \lambda n(z)$. Thus ϕ is the exponential map of the normal bundle of $\partial\Omega$, relative to the Euclidean norm on $\mathbb{R}^n$. In particular, ϕ defines a C^{r-1} diffeomorphism of some open neighbourhood of $\partial\Omega \times \{0\} \subset \partial\Omega \times \mathbb{R}$ onto an open neighbourhood U of $\partial\Omega$ in $\mathbb{R}^n$ (note that $n(\ )$ is only of class C^{r-1} on $\partial\Omega$). If $X \in U$ we may write $\phi^{-1}(X) = (\pi(X),\lambda(X))$, where $\pi: U \to \partial\Omega$, $\lambda: U \to \mathbb{R}$ are C^{r-1} functions. Thus

$$X - \pi(X) = \lambda(X)n(\pi(X)),\ X \in U.$$

Differentiating this expression with respect to X and evaluating at $y \in \mathbb{R}^n$ we see that

$$y - D\pi_X(y) = D\lambda_X(y)n(\pi(X)) + \lambda(X)Dn_{\pi(X)}(D\pi_X(y)) .$$

Take the inner product with $n(\pi(X))$ to obtain

$$<y,n(\pi(X))> = D\lambda_X(y) + \lambda(X)<Dn_{\pi(X)}(D\pi_X(y)),n(\pi(X))> .$$

Now $<Dn_{\pi(X)}(D\pi_X(y)),n(\pi(X))> = 0$ as is seen by differentiating the identity $<n(\pi(X)),n(\pi(X))> = 1$. Hence

$$D\lambda_X(y) = <y,n(\pi(X))> .$$

But $<y,n(\pi(X))>$ is a C^{r-1} function of X. Therefore, $D\lambda$ is C^{r-1} and so λ is C^r. Since $d(x) = \lambda(x)$, the result follows. □

Continuing with the notation of the proof of Lemma 2.5.14, we see that $Dd_x = <\cdot,n(\pi(X))>$, for $x \in \partial\Omega$ and so Dd is non-vanishing on $\partial\Omega$. Hence $-d$ is a continuous defining function for Ω which is C^r on some neighbourhood of $\partial\Omega$ in $\mathbb{R}^n$.

Now suppose Ω is a bounded domain in $\mathbb{C}^n$ with C^2 boundary. As in Lemma 2.5.14, we let $d: \mathbb{C}^n \to \mathbb{R}$ denote the signed Euclidean distance function to the boundary of Ω.

Proposition 2.5.15. Let Ω be a bounded domain in $\mathbb{C}^n$ with C^2 boundary and suppose that $d: \mathbb{C}^n \to \mathbb{R}$ is C^2 on the open neighbourhood U of $\partial\Omega$ in $\mathbb{C}^n$. Then the following conditions are equivalent:

i) Ω is Levi pseudoconvex.

ii) The Levi form of $-\log(d)$ is positive semi-definite on $U \cap \Omega$.

Proof. First suppose that $L(-\log d)$ is positive semi-definite on $U \cap \Omega$. Set $\delta = -\log d = \log d^{-1}$. Computing we find that

$$\frac{\partial^2\delta}{\partial z_i\partial\bar{z}_j} = -d^{-1}\frac{\partial^2 d}{\partial z_i\partial\bar{z}_j} + d^{-2}\frac{\partial d}{\partial z_i}\overline{\frac{\partial d}{\partial z_j}} \quad \text{on } U .$$

Hence for $z \in U \cap \Omega$, we have

$$\sum_{i,j} \frac{\partial^2 d}{\partial z_i \partial \bar{z}_j}(z) Z_i \bar{Z}_j \geq 0, \text{ provided that } \sum_i \frac{\partial d}{\partial z_i}(z) Z_i = 0 .$$

By continuity, this result holds on $\partial\Omega$ and so Ω is Levi pseudoconvex.

For the converse, we first note that since $-d$ is a C^2 defining function for Ω, at least on U, Lemma 2.5.10 implies that $L(-d)$ is positive semi-definite on holomorphic tangent vectors to $\partial\Omega$. Suppose that there exists $z \in U \cap \Omega$ such that the Levi form of $-\log d$ is not positive semi-definite at z. This implies that there exists a unit vector $u \in \mathbb{C}^n$ such that

$$\frac{\partial^2}{\partial t \partial \bar{t}} (\log d(z+tu))_{t=0} > 0 .$$

As in the proof of Theroem 2.5.13, Taylor's theorem implies that

$$\log d(z+tu) = \log d(z) + \mathrm{Re}(at+bt^2) + c|t|^2 + 0(|t|^2) ,$$

for constants $a,b \in \mathbb{C}$. Choose $\zeta \in \mathbb{C}^n$ such that $z + \zeta \in \partial\Omega$ and $d(z) = \|\zeta\|$. Consider the holomorphic curve

$$\phi(t) = z + tu + \zeta \exp(at+bt^2)$$

and note that $\phi(0) = z + \zeta \in \partial\Omega$. By the triangle inequality we have

$$d(\phi(t)) \geq d(z+tu) - \|\zeta\| \exp(at+bt^2), \ t \in \mathbb{C} .$$

Using the expression above for $\log d(z+tu)$ it follows that for sufficiently small values of $|t|$ we have

$$d(\phi(t)) \geq \|\zeta\| (\exp(c|t|^2/2) - 1) |\exp(at+bt^2)| .$$

But this estimate implies that at $t = 0$ we have

$$\frac{\partial}{\partial t} d(\phi(t)) = 0 \text{ and } \frac{\partial^2}{\partial t \partial \bar{t}} d(\phi(t)) > 0$$

and so

$$\sum_i \frac{\partial d}{\partial z_i}(z+\zeta)\phi_i'(0) = 0, \quad \sum_{i,j} \frac{\partial^2 d}{\partial z_i \partial \bar{z}_j}(z+\zeta)\phi_i'(0)\overline{\phi_j'(0)} < 0 ,$$

contradicting the positive semi-definiteness of L(d) on holomorphic tangent vectors to $\partial\Omega$. □

Definition 2.5.16. Let Ω be a domain in $\mathbb{C}^n$ and $\phi: \Omega \to \mathbb{R}$ be C^2. We say that ϕ is *plurisubharmonic* (psh for short) if $L(\phi)$ is positive semi-definite on Ω. We say ϕ is *strictly plurisubharmonic* if $L(\phi)$ is positive definite on ∂.

Remark. Notice that if $\phi: \Omega \to \mathbb{R}$ is C^2 and psh (respectively, strictly psh) and $h: \Omega' \to \Omega$ is biholomorphic then h is psh (respectively, strictly psh).

In practice, it is useful to remove the differentiability requirement on psh functions. Thus suppose $\phi: \Omega \to [-\infty,+\infty)$ is semi-continuous from above. We say ϕ is psh if for arbitrary, $z, w \in \mathbb{C}^n$, the function $t \mapsto \phi(z+tw)$ is *subharmonic* wherever defined. We recall that a function u on a domain $U \subset \mathbb{C}$ is said to be subharmonic if given any compact subset K of U and continuous function h on K which is harmonic on $\overset{o}{K}$ and $\geq u$ on ∂K then $u \leq h$ on K. Equivalently, ϕ is psh if $L(\phi)$ is positive semi-definite in the distributional sense. That is, if $\int_\Omega L(u)(x)(X)\phi(x)d\lambda(x) \geq 0$ for all $u \in C_c^\infty(\Omega)$, $X \in \mathbb{C}^n$. We refer the reader to Hörmander [1] or Vladimirov [1] for the basic properties of subharmonic and psh functions that are used in several complex variables. We should stress that the only reference to subharmonic functions in these notes will be in this section.

The next result is a generalisation of Proposition 2.5.15 to domains which do not have smooth boundary.

Proposition 2.5.17. If Ω is a domain of holomorphy in $\mathbb{C}^n$ then $-\log d(x)$ is psh.

Proof. See Bremermann [2], Hörmander [1] or Vladimirov [1]. □

Definition 2.5.18. A domain Ω in $\mathbb{C}^n$ is said to be *pseudoconvex* if $-\log d(x)$ is psh.

We now define a very strong form of pseudoconvexity.

Definition 2.5.19. A domain Ω in $\mathbb{C}^n$ is said to be *0-complete* or *holomorphically complete* if there exists $\phi \in C^\infty_{\mathbb{R}}(\Omega)$ such that

1. ϕ is strictly psh.

2. For every $a \in \mathbb{R}$, $\Omega_a = \{z \in \Omega: \phi(z) < a\}$ is relatively compact in Ω.

It is not hard to show that a domain Ω in $\mathbb{C}^n$ is pseudoconvex if and only if it is 0-complete. See, for example, Hörmander [1] or Vladimirov [1]. We shall prove directly that every domain of holomorphy is 0-complete.

Theorem 2.5.20. Let Ω be a domain of holomorphy. Then Ω is 0-complete.

Proof. Since Ω is holomorphically convex we may find a normal exhaustion $(K_n)_{n\geq 1}$ of Ω. Choose an increasing sequence (U_n) of relatively compact open subsets of Ω such that for all n, $K_n \subset U_n$ and U_n is relatively compact in U_{n+1}. Since $\hat{K}_n = K_n$ and $\bar{U}_{n+1} \setminus U_n$ is compact there exist $f_{nk} \in A(\Omega)$, $1 = 1,\ldots,k(n)$, such that

$$\|f_{nk}\|_{K_n} < 1 \text{ and } \max_k |f_{nk}(z)| > 1,\ z \in \bar{U}_{n+1} \setminus U_n .$$

Raising the f_{nk} to sufficiently high powers we may further assume that

A.... $$\sum_{n=1}^{k(n)} |f_{nk}(z)|^2 < 2^{-n},\ z \in K_n$$

B.... $$\sum_{k=1}^{k(n)} |f_{nk}(z)|^2 > n,\ z \in \bar{U}_{n+1} \setminus U_n .$$

Since $\Omega \subset \mathbb{C}^n$, we may further require that for every $z \in K_n$ there exist n of the f_{nk} which gives local holomorphic coordinates at z. Define

$$\phi(z) = \sum_{n,k} |f_{nk}(z)|^2,\ z \in \Omega.$$

The sum converges by estimate A for all $z \in \Omega$. We claim that $\phi \in C^\infty_{\mathbb{R}}(\Omega)$. This follows since $\Sigma f_{nk}(z)\overline{f_{nk}(\zeta)}$ is uniformly convergent on compact subsets of $\Omega \times \Omega$ and so, by Corollary 2.1.8, the sum is analytic in z and $\bar{\zeta}$. Next notice that estimate B implies that $\phi(z) > n$, for $z \in \Omega \setminus U_n$. Hence for all $a \in \mathbb{R}$, $\Omega_a = \{z: \phi(z) < a\}$ is relatively compact. Finally, we have

$$L(\phi) = \sum \frac{\partial f_{nk}}{\partial z_i} \overline{\frac{\partial f_{nk}}{\partial z_j}}$$

and so $L(\phi)$ is certainly positive semi-definite. To prove positive definitiness suppose that $L(\phi)(z)(Z) = 0$, $z \in K_n$. This implies that $\sum_{i,k} \frac{\partial f_{nk}}{\partial z_i}(z)Z_i = 0$. But it is possible to find n of the f_{nk} which give a local coordinate system at z. Hence $Z = 0$. □

Remarks.

1. We call functions ϕ that satisfy the conditions of Definition 2.5.19 *strictly psh exhaustion functions for* Ω. Notice that our definition of 0-completeness is invariant under biholomorphic maps and makes no explicit mention of the boundary of Ω. Later, in Chapter 5, we shall generalise our 0-completeness definition to complex manifolds.

2. It follows easily from Sard's theorem that if Ω is 0-complete then Ω is the limit of an increasing sequence of s.L.p. domains (see also the discussion in §4, Chapter 7).

Next we wish to give a brief discussion of some "continuity" principles that hold for pseudoconvex domains. Motivation for the next definition is given by the proof of Theorem 2.5.13.

Definition 2.5.21. A domain Ω in $\mathbb{C}^n$ is said to satisfy the *weak continuity principle* if, given any sequence $\{V_n\}$ of holomorphically embedded discs in Ω which satisfy

1. For all n, $V_n \cup \partial V_n$ is a relatively compact subset of Ω.
2. $\lim V_n = V_0$ exists and ∂V_0 is relatively compact subset of Ω.

Then, provided that V_0 is bounded, V_0 is a relatively compact subset of Ω.

Using subharmonic functions it is not too hard to show that a domain is pseudoconvex if and only if it satisfies the weak continuity principle (see Bremermann [2], Vladimirov [1]). If we say that a domain Ω in $\mathbb{C}^n$ is H-pseudoconvex if it contains all its generalised Hartogs figures, then it is straightforward to show that Ω is H-pseudoconvex if and only if it satisfies the weak continuity principle (see Grauert and Fritzsche [1] and also Andreotti and Grauert [1; page 217ff]).

Let us summarise the relations between the various definitions we have been considering.

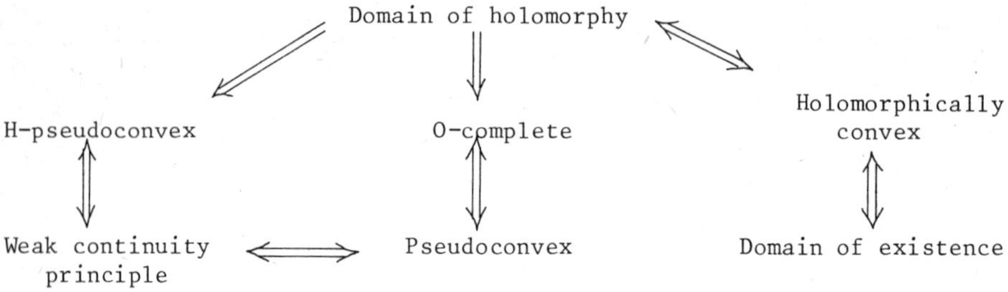

In Chapter 12 we shall prove the fundmamental result that every O-complete domain in $\mathbb{C}^n$ is a domain of holomorphy. As a consequence, the definitions above will have all been shown to be equivalent.

Finally the reader interested in pseudoconvexivity and its relationship with domains of holomorphy should certainly consult the original works of Oka [3] and the H. Cartan seminar [2].

Exercises.

1. Show that Lemma 2.5.10 is false if $\partial\Omega$ is only C^1. Can you find an example of a domain in $\mathbb{C}^n$ with C^1 boundary such that $d(x)$ is not differentiable at any point of $\mathbb{C}^n$?

2. Show

   a) The interior of an arbitrary intersection of pseudoconvex domains is pseudoconvex (cf. Example 15, §4).

   b) If Ω is a domain in $\mathbb{C}^n$ such that every point $z \in \partial\Omega$ has an open neighbourhood U such that $U \cap \Omega$ is pseudoconvex then Ω is pseudoconvex (the same result is true for domains of holomorphy but depends on the equivalence between domains of holomorphy and pseudoconvex domains).

   (Both a) and b) use elementary properties of subharmonic functions).

3. (Kohn and Nirenberg [1]). Show that the domain Ω in $\mathbb{C}^2$ defined by $\text{Re}(w) + |zw|^2 + |z|^8 + \frac{15}{7}|z|^2\text{Re}(z^6) < 0$ is L.p. and s.L.p. at every point of $\partial\Omega$ except 0.

(Ω is not convex at 0 in any local system of holomorphic coordinates, See Kohn and Nirenberg [1] and also Y.T. Siu [1]).

§6. The Bergman kernel function

In this seciton we describe how some domains of holomorphy in $\mathbb{C}^n$ have an intrinsic strictly psh exhaustion function (granted the Euclidean metric structure on $\mathbb{C}^n$). For the most part we either omit or give very brief indications of proofs. Full details may be found in Bergman [1,2] or Fuks [1]. We shall also assume some elementary Hilbert space theory.

Throughout this section we shall suppose that Ω is a bounded domain in $\mathbb{C}^n$. We have already shown in Exercise 2, §1, that $L^2(\Omega)$ is a Hilbert space with inner product defined by $(f,g) = \int_\Omega f\bar{g}d\lambda$, where $d\lambda$ is Lebesgue measure on $\mathbb{C}^n$. We set $|f| = (f,f)^{\frac{1}{2}}$, $f \in L^2(\Omega)$. Since $L^2(\Omega)$ is separable, we may find a countable orthonormal (Hilbert space) basis $\{\phi_j : j \geq 1\}$ for $L^2(\Omega)$. We remark that if $\{\psi_j\}$ is an arbitrary Hilbert space basis for $L^2(\Omega)$ then we can construct an orthonormal basis for $L^2(\Omega)$ from $\{\psi_j\}$ using the Gram-Schmidt orthogonalisation process.

Proposition 2.6.1. Let $\{\phi_j\}$ be an orthonormal basis for $L^2(\Omega)$. Given $f \in L^2(\Omega)$, set $a_j = (\phi_j, f)$, $j \geq 1$. Then

1. $f = \sum_{j=1}^{\infty} a_j\phi_j$, where convergence is uniform on compact subsets of $\partial\Omega$.

2. $|f|^2 = \sum_{j=1}^{\infty} |a_j|^2$ (Parseval's equality).

Proof. Straightforward and based on the estimates of Exercise 2, §1. See also the proof of Proposition 2.6.2 below. □

Proposition 2.6.2. Let $\{\phi_j\}$ be an orthonormal basis for $L^2(\Omega)$. Then the series

$$\sum_{j=1}^{\infty} \phi_j(z)\overline{\phi_j(\zeta)}$$

converges uniformly on compact subsets of $\Omega \times \Omega$ to a function $K_\Omega(z,\bar{\zeta})$ which is analytic in $(z,\bar{\zeta})$. In particular, K_Ω is C^∞.

Proof. Let $D_0 = D(z_0;r)$, $D_1 = D(\zeta_0;s)$ be relatively compact polydiscs in Ω. By Exercise 2, §1, there exists $C_0 \geq 0$ such that for all $f \in L^2(\Omega)$, $\|f\|_{D_0} \leq C_0|f|$. Therefore for $z \in D_0$ we have

$$\sum_{j=1}^{m} |\phi_j(z)|^2 = \int_\Omega |\sum_{j=1}^{m} \phi_j(\tau)\overline{\phi_j(z)}|^2 d\lambda(\tau)$$

$$\geq C_0 \|\sum_{j=1}^{m} \phi_j(\tau)\overline{\phi_j(z)}\|^2_{D_0}$$

$$\geq C_0 \|\sum_{j=1}^{m} |\phi_j(z)|^2\|^2_{D_0} .$$

Hence

$$\sum_{j=1}^{m} |\phi_j(z)|^2 \leq C_0^{-1}, \quad z \in D_0 .$$

Similarly,

$$\sum_{j=1}^{m} |\phi_j(\zeta)|^2 \leq C_1^{-1}, \quad \zeta \in D_1 .$$

By the Cauchy-Schwarz inequality

$$\left(\sum_{j=1}^{m} |\phi_j(z)\overline{\phi_j(\zeta)}|\right)^2 \leq \left(\sum_{j=1}^{m} |\phi_j(z)|^2\right)^2\left(\sum_{j=1}^{m} |\phi_j(\zeta)|^2\right)^2$$

and so

$$\sum_{j=1}^{m} |\phi_j(z)\overline{\phi_j(\zeta)}| \leq (C_0C_1)^{-1}, \quad z \in D_0, \ \zeta \in D_1 .$$

Therefore $\sum_{j=1}^{\infty} \phi_j(z)\overline{\phi_j(\zeta)}$ converges uniformly on compact subsets of $\Omega \times \Omega$ and so the result follows by Corollary 2.1.8. □

Definition 2.6.3. Let $\{\phi_j\}$ be an orthonormal basis for $L^2(\Omega)$. Then the function $K_\Omega(z,\bar{\zeta}) = \sum_{j=1}^{\infty} \phi_j(z)\overline{\phi_j(\zeta)}$ is called the *(Bergman) kernel function* of the domain Ω, relative to the orthonormal basis $\{\phi_j\}$.

Theorem 2.6.4. The kernel function $K_\Omega(z,\bar{\zeta})$ is independent of the choice of orthonormal basis for $L^2(\Omega)$. It satisfies the following characteristic variational property:

Given $z_0 \in \Omega$, the function

$$f(z) = K_\Omega(z,\bar{z}_0)/K_\Omega(z_0,\bar{z}_0)$$

is the unique function in $L^2(\Omega)$ which minimises the integral $|f|$, subject to the normalising condition $f(z_0) = 1$.

Proof. Straightforward and we refer to the references. □

In view of Theorem 2.6.4 we may now talk about *the* kernel function of the domain Ω.

Theorem 2.6.5. (Reproducing property of the kernel function). For all $f \in L^2(\Omega)$, we have

$$f(z) = \int_\Omega f(\zeta)K_\Omega(z,\bar{\zeta})d\lambda(\zeta).$$

Proof. Take the Fourier series expansion of f relative to an orthonormal basis of $L^2(\Omega)$ and integrate term by term. □

Example. Set $D = D(0;r_1,\ldots,r_n)$, $E = E(0;r)$. Then $\left\{\frac{z^m}{m!}: m \in \mathbb{N}^n\right\}$ is easily seen to be an orthogonal basis for both $L^2(D)$ and $L^2(E)$. Normalising and computing the kernel functions we find

$$K_D(z,\bar{\zeta}) = \pi^{-n}r_1^2\ldots r_n^2 \prod_{j=1}^n (r_j - z_j\bar{\zeta}_j)^{-2}$$

$$K_E(z,\bar{\zeta}) = \pi^{-n}r^2 n!\left(r^2 - \sum_{j=1}^n z_j\bar{\zeta}_j\right)^{-n-1} .$$

Observe that $K_D(z,\bar{z})$, $K_E(z,\bar{z})$ are strictly psh exhaustion functions for their respective domains. The reader may find computations of the kernel functions of the "classical domains" (see Chapter 4, §2) in Hua [1].

Next we examine how the kernel function transforms under biholomorphic maps.

Proposition 2.6.6. Let $h: \Omega \to \Omega'$ be a biholomorphic map between bounded domains in $\mathbb{C}^n$. Then

$$K_\Omega(z,\bar{z}) = K_{\Omega'}(h(z),\overline{h(z)})\,|\det_{\mathbb{C}} Dh_z|^2 .$$

Proof. A straightforward computation using the change of variables formula for multiple integrals and the fact that $\det_{\mathbb{R}} Dh_z = |\det_{\mathbb{C}} Dh_z|^2$ (see §4, Chapter 5). □

Remarks.

1. Proposition 2.6.6 implies that the kernel function is invariantly defined as a section of an appropriate tensor bundle on Ω. We return to this point in Chapter 5.

3. Note that Proposition 2.6.6 in combination with the computations of the Example, shows that the open polydisc and Euclidean disc in $\mathbb{C}^n$ are biholomorphically inequivalent, $n > 1$. See also §2 of Chapter 4.

Corollary 2.6.7. The Levi form of $\log K_\Omega(z,\bar{z})$ is invariant under biholomorphic transformations.

Because of Corollary 2.6.7 and the analogy with pseudoconvexivity we prefer to work with $\log K_\Omega(z,\bar{z})$ rather than $K_\Omega(z,\bar{z})$.

Proposition 2.6.8. Let Ω be a bounded domain in $\mathbb{C}^n$. Then $\log K_\Omega(z,\bar{z})$ is a C^∞ strictly psh function.

Proof. A straightforward computation that makes use of the fact that since Ω is bounded we can define local coordinate systems at any point of Ω using elements of a fixed orthonormal basis for $L^2(\Omega)$. □

The question now arises as to the conditions under which $\log K_\Omega(z,\bar{z})$ gives a strictly psh exhaustion function for a domain of holomorphy Ω. In general it does not (see Bremermann [3]). However, it is certainly true that a sufficient condition for Ω to be a domain of holomorphy is that $\log K_\Omega(z,\bar{z})$ is a strictly psh exhaustion function for Ω. Moreover, it can be shown that every domain of holomorphy is the limit of an increasing sequence of domains of holomorphy for which $\log K(z,\bar{z})$ is a strictly psh exhaustion function (see Bremermann [2] and compare with the corresponding approximation of domains of holomorphy by s.L.p. domains). Finally we should mention that if we define a weighted inner product $(f,g)_\phi = \int_\Omega f\bar{g}e^{-\phi}d\lambda$ and let $L^2(\Omega;\phi)$ denote the corresponding Hilbert space of analytic functions on Ω, then the kernel theory continues to hold. Sharp estimates on the growth of the associated kernel function at the boundary of Ω are given in Hörmander [3]. See also Bremermann [2]. Notice that the introduction of weighting functions amounts to a change of metric on Ω.

For a survey of recent results on the kernel function and its applications to complex analysis see Diederich [1].

Exercises.

1. Verify that the kernel function of a domain Ω is Hermitian: $K_\Omega(z,\zeta) = \overline{K_\Omega(\zeta,z)}$, $z,\zeta \in \Omega$.

2. Verify the expressions for the kernel functions of the polydisc and Euclidean disc given in the example.

§7. The Cousin problems

In this section we wish to consider the problem of constructing meromorphic functions with specified principle parts or zero and pole sets on a given domain in $\mathbb{C}^n$, $n > 1$. These questions were first raised by P. Cousin around the turn of the century and, following H. Cartan [1], now bear his name. As yet, of course, we have not defined meromorphic functions of more than one variable (see Chapter 3) but, as we showed in Chapter 1, it is possible to formulate both the Mittag-Leffler and Weierstrass theorems in a way that avoids explicit mention of meromorphic functions. We adopt this approach here but we must caution the reader that our consequent definition of the Cousin problems is *not* the classical one (see the remarks below).

Cousin's problem A.

Let Ω be a domain in $\mathbb{C}^n$ and $\{U_i: i \in I\}$ be an open cover of Ω. Suppose we are given $f_{ij} \in A(U_{ij})$ such that for all i,j,k

$$f_{ij} = -f_{ji} \text{ on } U_{ij}; \quad f_{ij} + f_{jk} + f_{ki} = 0 \text{ on } U_{ijk} .$$

When can we find $f_i \in A(U_i)$ such that $f_{ij} = f_j - f_i$ on U_{ij} for all $i,j \in I$?

We say Ω is a *Cousin A domain* if we can always solve the Cousin A problem on Ω.

Cousin's problem B.

Let Ω be a domain in $\mathbb{C}^n$ and $\{U_i: i \in I\}$ be an open cover of Ω. Suppose we are given $f_{ij} \in A^*(U_{ij})$ such that for all i,j,k

$$f_{ij} = f_{ji}^{-1} \text{ on } U_{ij}; \quad f_{ij}f_{jk}f_{ki} = 1 \text{ on } U_{ijk} .$$

When can we find $f_i \in A(U_i)$ such that $f_{ij} = f_j/f_i$ on U_{ij} for all $i,j \in I$?

We say Ω is a *Cousin B domain* if we can always solve the Cousin B problem on Ω.

Proposition 2.7.1. Let Ω be a domain in $\mathbb{C}^n$. Then

1. Ω is a Cousin A domain if and only if we can solve the Cauchy-Riemann equations on Ω.

2. Ω is a Cousin B domain if and only if every holomorphic line bundle on Ω is holomorphically trivial.

Proof. To say that we can solve the Cauchy-Riemann equations on Ω means that if we are given $f_j \in C^\infty(\Omega)$, $1 \leq j \leq n$, such that $\partial f_j/\partial \bar{z}_i = \partial f_i/\partial \bar{z}_j$, $1 \leq i, j \leq n$, then there exists $u \in C^\infty(\Omega)$ such that $\partial u/\partial \bar{z}_j = f_j$, $1 \leq j \leq n$ (see §3, especially Theorem 2.3.1). Suppose that we can solve the Cauchy-Riemann equations on Ω. Then, exactly as in the proof of Theorem 1.3.2, we can solve the Cousin A problem on Ω. We defer the proof of the converse to Chapter 6. The second statement is an immediate consequence of the definition of a holomorphically trivial line bundle (see §5, Chapter 1). □

We know from Chapter 1 that every domain in $\mathbb{C}$ is a Cousin A and Cousin B domain. Notice that if Ω is a domain in $\mathbb{C}$ and $f_{ij} \in A(U_{ij})$ is the data for a Cousin A problem on Ω then we can find $m_j \in M(U_j)$ such that $f_{ij} = m_j - m_i$. Indeed, we can choose $m_j \in A(U_j)$ satisfying these conditions since Ω is a Cousin A domain! Hence, for domains in $\mathbb{C}$, there is really no difference between the Cousin A problem and the problem of constructing meromorphic functions with specified principal parts. Similarly for the multiplicative problem. It turns out that this equivalence between the Cousin A and B problems and the problem of finding meromorphic functions with specified principal parts or pole and zero sets no longer holds for domains in $\mathbb{C}^n$, $n > 1$. As we shall see in Chapter 12, it is possible, for example, to have domains in $\mathbb{C}^2$ for which we can always construct meromorphic functions with specified pole and zero sets but which are nevertheless not Cousin B (or A) domains. To clarify this point we now give the original definition of a Cousin II domain.

Definition 2.7.2. We say that a domain Ω in $\mathbb{C}^n$ is a *Cousin II domain* if given any open cover $\{U_i\}$ of Ω and analytic functions $f_i \in A^*(U_i)$ such that $f_i/f_j \in A^*(U_{ij})$ then there exists $F \in A(\Omega)$ such that $Ff_i^{-1} \in A^*(U_i)$ for all i.

Remark. We shall see in §4 of Chapter 3, if Ω is a Cousin II domain then we can construct meromorphic functions on Ω with specified pole and zero sets. However, a Cousin II domain need not be a Cousin B domain if $n > 1$.

Examples.

1. Every open polydisc $D \subset \mathbb{C}^n$ is a Cousin A domain. To prove this we shall show that the Cauchy-Riemann equations are solvable in D. Our proof will be very close to that of Theorem 1.3.1. Let $D = D(0;r_1,\ldots,r_n)$ and for $s \in (0,1]$, set $D_s = sD$. Suppose $f \in A(D)$. By Theorem 2.1.5

$$f(z) = \sum \frac{1}{m!} \partial^m f(0) z^m, \quad z \in D,$$

with uniform convergence on compact subsets of D. Hence, given $s \in (0,1)$ and $\varepsilon > 0$, there exists $N > 0$ such that $\|f - f_N\|_{\bar{D}_s} \leq \varepsilon$, where $f_N(z) = \sum_{|m|\leq N} \frac{\partial^m f(0)}{m!} z^m$. Hence polynomials are dense in A(D). Using this approximation result together with Theorem 2.3.1 we can now use the method of proof of Theorem A1.8 (Theorem 1.3.1) to show that the Cauchy-Riemann equations are solvable on D. We leave details to the reader (see also Theorem 5.8.2).

This example can be generalised in the following way: We say that $\Omega \subset \mathbb{C}^n$ is a *Runge domain* if polynomials are dense in $A(\Omega)$. It may be proved, using an ingenious inductive argument due to Oka, that the Cauchy-Riemann equations may always be solved on a Runge domain. For further details and references we refer the reader to Hörmander [1], Gunning and Rossi [1].

2. The domain $\Omega = \mathbb{C}^2 \setminus \{0\}$ is not a Cousin A domain. We shall show that the Cauchy-Riemann equations are not always solvable on Ω. Let $z = (z_1,z_2) \in \Omega$ and set $\|z\|^2 = |z_1|^2 + |z_2|^2$. We define functions f_1, f_2 by

$$f_1(z) = \frac{\partial}{\partial \bar{z}_1}(\bar{z}_2/z_1\|z\|^2),\ z_1 \neq 0$$
$$= \frac{\partial}{\partial \bar{z}_1}(\bar{z}_1/z_2\|z\|^2),\ z_2 \neq 0$$

$$f_2(z) = \frac{\partial}{\partial \bar{z}_2}(\bar{z}_2/z_1\|z\|^2),\ z_1 \neq 0$$
$$= \frac{\partial}{\partial \bar{z}_2}(\bar{z}_1/z_2\|z\|^2),\ z_2 \neq 0 \quad .$$

The identity $\frac{1}{z_1 z_2} = \bar{z}_2/z_1\|z\|^2 + \bar{z}_1/z_2\|z\|^2$, $z_1z_2 \neq 0$, implies that $f_1, f_2 \in C^\infty(\Omega)$. Clearly we also have $\partial f_1/\partial \bar{z}_2 = \partial f_2/\partial \bar{z}_1$ on Ω. Suppose that there exists $u \in C^\infty(\Omega)$ such that $\partial u/\partial \bar{z}_j = f_j$, $j = 1,2$. The function $G = z_1 u - \bar{z}_2/\|z\|^2$ is holomorphic for $z_1 \neq 0$ since $z_1^{-1}\partial G/\partial \bar{z}_j = \partial u/\partial \bar{z}_j - f_j = 0$, $j = 1,2$. Now G is locally bounded on Ω and so, by the Riemann removable singularities theorem, G extends to an analytic function on Ω. Hartogs theorem implies that G extends to an analytic function on $\mathbb{C}^2$. But $G(0,z_2) = 1/z_2$, $z_2 \neq 0$, and so G cannot be holomorphic at 0. Contradiction and so we cannot solve this set of Cauchy-Riemann equations on Ω. We shall show in Chapter 12 that $\mathbb{C}^n \setminus \{0\}$ is a Cousin A domain for $n \geq 3$.

Proposition 2.7.3. Let Ω be a domain in $\mathbb{C}^n$. Then Ω is a Cousin B domain if and only if we can solve the Cauchy-Riemann equations on Ω and $H^2(\Omega, \mathbb{Z}) = 0$.

Proof. If we can solve the Cauchy-Riemann equations on Ω and $H^2(\Omega, \mathbb{Z}) = 0$ then, exactly as in the Proof of Theorem 1.3.2 we can solve the Cousin B problem on Ω. We defer the proof of the converse to Chapter 6. □

Example 3. Every open polydisc in $\mathbb{C}^n$ is a Cousin B domain.

The next theorem is very deep and we defer the proof until Chapter 12.

Theorem. The Cauchy-Riemann equations are solvable on every domain of holomorphy.

Corollary. If Ω is a domain of holomorphy, Ω is a Cousin A domain. If, in addition, $H^2(\Omega, \mathbb{Z}) = 0$, Ω is a Cousin B domain.

As we have already seen there are Cousin A domains which are not domains of holomorphy. However, in case $\Omega \subset \mathbb{C}^2$ it is true that a Cousin A(B) domain is a domain of Holomorphy (see H. Cartan [2] and Chapter 12). To actually characterise domains of holomorphy in terms of Cauchy-Riemann equations we have to take account of higher order versions of the Cauchy-Riemann equations. See Chapter 12.

Exercise. Show that Cauchy-Riemann equations are solvable on the Euclidean disc $E(r)$ (Use Exercise 1, §1).

CHAPTER 3. LOCAL RINGS OF ANALYTIC FUNCTIONS

Introduction

In this chapter we shall make a study of algebraic and analytic properties of the ring $\mathbb{C}\{z_1 - a_1,\dots,z_n - a_n\}$ of convergent power series at a point $(a_1,\dots,a_n) \in \mathbb{C}^n$. It turns out that there is a rich and deep relationship between algebraic properties of power series rings and local properties of analytic functions and analytic sets. This relationship is expressed in its most elegant and powerful form using the concept of "coherence" which we introduce later in Chapter 7. In this chapter we show how to define meromorphic functions of more than one complex variable. We also prove some elementary facts about local properties of analytic sets and some not so elementary results about modules over $\mathbb{C}\{z_1 - a_1,\dots,z_n - a_n\}$. The main technical result we need is the most important Weierstrass Preparation theorem which we prove in section 2.

1. Elementary properties of power series rings

Throughout this Chapter $\mathbb{C}\{z_1 - a_1,\dots,z_n - a_n\}$ or just $\mathbb{C}\{z - a\}$ will denote the ring of convergent power series at a point $(a_1,\dots,a_n) = a \in \mathbb{C}^n$. We let $\mathcal{U}_a$ denote the set of all open neighbourhoods of a point $a \in \mathbb{C}^n$.

We first remark that $\mathbb{C}\{z - a\}$ has the structure of a commutative ring with identity under the obvious definitions of multiplication and addition. Given $a,b \in \mathbb{C}^n$, there is a natural isomorphism between $\mathbb{C}\{z - a\}$ and $\mathbb{C}\{z - b\}$ defined by mapping $\Sigma\alpha_m(z-a)^m$ to $\Sigma\alpha_m(z - b)^m$. Because of this isomorphism we may safely take $a = 0$ and this will result in no loss of generality as far as algebraic properties of these rings are concerned.

We now show another way to construct the ring $\mathbb{C}\{z - a\}$ in a "coordinate-free" manner. We define an equivalence relation on the set of all analytic functions which are defined on some neighbourhood of a. Suppose $f \in A(U)$, $g \in A(V)$, $U,V \in \mathcal{U}_a$. We write $f \sim_a g$ if there exists $W \in \mathcal{U}_a$ such that $f|W = g|W$. That is, $f \sim_a g$ if and only if $f = g$ on some neighbourhood of a. Clearly $\sim_a$ is an equivalence relation and we let $\mathcal{O}_a$ denote the set of equivalence classes of $\mathcal{O}_a$ ("$\mathcal{O}$" in honour of the Japanese mathematician K. Oka who proved many of the fundamental results in the theory of several complex variables). If $f \in A(U)$, $U \in \mathcal{U}_a$, we let f_a denote the $\sim_a$ equivalence class of f in $\mathcal{O}_a$. We call f_a the *germ* of f at a. Obviously, $\mathcal{O}_a$ inherits the structure of a commutative ring with 1 from the corresponding structures on the A(U). We claim that $\mathbb{C}\{z-a\}$ and $\mathcal{O}_a$ are naturally isomorphic. Indeed, if $P \in \mathbb{C}\{z-a\}$, we may define $\theta(P) \in \mathcal{O}_a$ to be the germ at a of the analytic function defined on some neighbourhood of a by the power series P. Conversely, if $f_a \in \mathcal{O}_a$ is the germ of an analytic function $f \in A(U)$, $U \in \mathcal{U}_a$, we define $\theta^{-1}(f_a)$ to be the convergent power series of f at a. Clearly θ is a ring isomorphism. In the sequel, we identify the rings $\mathbb{C}\{z-a\}$, $\mathcal{O}_a$ via θ and generally use the more abbreviated notation $\mathcal{O}_a$ in preference to $\mathbb{C}\{z-a\}$. We generally omit reference to n. If essential we write ${}_n\mathcal{O}_a$ to denote the ring of germs of analytic functions at a point $a \in \mathbb{C}^n$.

Remark. Our construction of $\mathcal{O}_a$ as the ring of germs of analytic functions at a works equally well for other classes of functions. For example, we may construct the ring C_a of *germs of continuous functions* at a or the ring $\mathcal{D}_a$ of *germs of* C^∞ *functions* at a. Clearly $\mathcal{O}_a \subset \mathcal{D}_a \subset C_a$. However, the rings C_a and $\mathcal{D}_a$ do not have a simple representation in terms of power series rings nor do they have the rich algebraic structure of $\mathcal{O}_a$ and most of the results we prove for $\mathcal{O}_a$ fail for C_a and $\mathcal{D}_a$.

Lemma 3.1.1. The ring $\mathcal{O}_a$ is an integral domain. That is, if $f,g \in \mathcal{O}_a$ and $fg = 0$, then either $f = 0$ or $g = 0$.

Proof. We may choose a connected open neighbourhood U of a and $F,G \in A(U)$ such that $F_a = f$ and $G_a = g$. Now $fg = 0$ implies that $FG \equiv 0$ on U by uniqueness of analytic continuation. Suppose $F \not\equiv 0$. Then F is non-zero on a non-empty subset V of U. Hence G is zero on V and, by uniqueness of analytic continuation, identically zero on U. Hence $g = 0$. □

Before giving the next lemma we recall

Definition 3.1.2. Let R be an integral domain with identity. An element $u \in R$ is said to be a *unit* if there exists $u^{-1} \in R$ such that $uu^{-1} = 1$.

Lemma 3.1.3.

1. An element $u \in \mathcal{O}_a$ is a unit if and only if $u(a) \neq 0$.
2. $\mathcal{O}_a$ has a unique maximal ideal m_a which is characterised as being the set of non-units of $\mathcal{O}_a$. That is, $f \in m_a$ if and only if $f(a) = 0$.

Proof. We first remark that if $f \in \mathcal{O}_a$ then $f(a)$ is, of course, just the value of any representative function for f at a; equivalently, the constant term in the power series for f. The lemma follows from the observations that if f is an analytic function which is non-zero at a, then $1/f$ is defined and analytic on some neighbourhood of a and any ideal of $\mathcal{O}_a$ which contains a unit is equal to $\mathcal{O}_a$. □

Exercises

1. Let U be an open subset of $\mathbb{C}^n$. Prove that A(U) is an integral domain if and only if U is connected.

2. Show that $\mathcal{D}_a$, the ring of germs of C^∞ complex valued functions at a point $a \in \mathbb{C}^n$, has a unique maximal ideal. Show also that $\mathcal{D}_a$ is not an integral domain.

3. Prove that $\mathcal{O}_a/m_a^{k+1}$ is naturally isomorphic to the complex vector space of complex valued complex polynomials on $\mathbb{C}^n$ of degree $\leq k$. (Hint: Look at the Taylor polynomial of degree k of an analytic function).

4. If m_a^∞ denotes the maximal ideal of $\mathcal{D}_a$, prove that $\mathcal{D}_a/(m_a^\infty)^{k+1}$ is naturally isomorphic to the vector space of complex valued real polynomials on $\mathbb{C}^n$ of degree $\leq k$.

5. Show that there is a natural isomorphism of m_a^k/m_a^{k+1} onto the complex vector space of complex valued homogeneous polynomials on $\mathbb{C}^n$ of degree k.

§2. Weierstrass Division and Preparation Theorems

In this section we prove the basic technical result which allows us to prove in §3 that $\mathcal{O}_a$ is a Noetherian unique factorization domain. Before giving the main result, we remark that we take coordinates $z = (z', z_n) = (z_1, \ldots, z_{n-1}, z_n)$ on $\mathbb{C}^n$, always using the prime to denote the $\mathbb{C}^{n-1}$ variable.

Theorem 3.2.1. (The Weierstrass Division Theorem) Let f be analytic on a neighbourhood Ω of $0 \in \mathbb{C}^n$ and suppose that $f(0, \ldots, 0, z_n)$ has a zero of multiplicity $p \geq 0$ at $z_n = 0$. Then we may find an open polydisc neighbourhood $D \subset \Omega$ of 0 such that every bounded analytic function g on D may be written uniquely in the form

$$g = qf + h,$$

where q and h are bounded analytic functions on D and h is a polynomial in z_n of degree $\leq p - 1$ with coefficients depending analytically on z'. q and h depend continuously on g in the sense that there exists a constant M, independent of g, such that

$$\|q\|_D, \ \|h\|_D \leq M\|g\|_D.$$

Proof. Choose a polydisc neighbourhood ω of 0, contained in Ω, on which f is bounded. On ω we may write

$$f(z) = \sum_{j=0}^{\infty} f_j(z') z_n^j$$

and, as is easily seen using Cauchy's inequalities (Corollary 2.1.7), each f_j is bounded on ω. Our hypotheses on f imply that $f_p(0) \neq 0$ and $f_j(0) = 0$, $j < p$. From now on we shall always work inside ω. Choose d,s > 0 such that if

$$D' = \{z' \in \mathbb{C}^{n-1} : |z_j| < d, \ 1 \leq j \leq n-1\}$$
$$D = \{z \in \mathbb{C}^n : z' \in D' \text{ and } |z_n| < s\}$$

then f_p is nonvanishing on D' and $1/f_p$ is bounded on D'. Multiplying f by $1/f_p$, it is no loss of generality to suppose that $f_p \equiv 1$ on D'.

Let $A^B(D)$ denote the algebra of bounded analytic functions on D. Define $\|u\| = \|u\|_D = \sup_D |u(z)|$, $u \in A^B(D)$. Then $A^B(D)$ is complete with respect to $\| \ \|$ and so has the structure of a Banach algebra.

Since $f_j(0) = 0$, $j < p$, there exists $C_1 \geq 0$ such that $\|f_j(z')z_n^j\| \leq C_1 \frac{ds^j}{p}$, $j < p$. Here C_1 is independent of d and s and depends only on, say, $\|f_j\|_\omega$. Similarly there exists $C_2 \geq 0$, independent of d and s, such that

$$\|\sum_{j>p} f_j(z')z_n^j\| \leq C_2 s^{p+1}$$

Hence

$$\|f - z_n^p\| \leq C_1 d(1 + s^p) + C_2 s^{p+1}$$

Now choose d and s sufficiently small so that

$$\|f - z_n^p\| \leq s^p/(2(p+1)) \qquad \ldots\ldots\; 1$$

From now on d,s and the corresponding polydiscs D',D will remain fixed. We now prove the division theorem in the special, trivial, case $f = z_n^p$. We derive the general case by regarding f as a pertubation of z_n^p, using estimate 1 above.

Given $g \in A^B(D)$, write

$$g = q(g)z_n^p + h(g),$$

where $q(g), h(g) \in A^B(D)$ and h(g) is a polynomial in z_n of degree $\leq p-1$. q(g) and h(g) are unique. Moreover

$$h(g)(z) = \sum_{j=0}^{p-1} \frac{\partial^j g}{\partial z_n^j}(z',0)\frac{z_n^j}{j!}$$

and so it follows from Cauchy's inequalities (Corollary 2.1.7) that

$$\|h(g)\| \leq p\|g\| \qquad \ldots.2$$

Hence, by the triangle inequality, $\|q(g)z_n^p\| \leq (p+1)\|g\|$ and so by Schwarz' Lemma (exercise 7, §1, Chapter 1)

$$\|q(g)\| \leq (p+1)s^{-p}\|g\| \qquad \ldots.3$$

(Equivalently, argue using the maximum modulus theorem on polydiscs εD, $0 < \varepsilon < 1$, and let $\varepsilon \to 1$).

Estimates 2 and 3 imply that the linear endomorphisms $g \mapsto h(g), q(g)$ of $A^B(D)$ are continuous.

We define a continuous linear map $A: A^B(D) \to A^B(D)$ by

$$A(u) = q(u)f + h(u) .$$

Now

$$\begin{aligned} \|(A-I)u\| &= \|q(u)f + h(u) - u\| \\ &= \|q(u)(f - z_n^p)\| \\ &\leq \|q(u)\|\|f - z_n^p\| \\ &\leq \tfrac{1}{2}\|u\| , \end{aligned}$$

using estimates 1 and 3. Hence $\|A - I\| \leq \frac{1}{2}$ and so A is a linear isomorphism (for a proof of this elementary result from Banach space theory see Dieudonné [1; page 148] or Field [1; page 189]). It follows that for any $g \in A^B(D)$, there exists $u \in A^B(D)$ such that

$$g = A(u) = q(u)f + h(u) .$$

This proves the first part of the theorem. Now $q(u) = q(A^{-1}(g))$ and so by estimate 3 we have

$$\begin{aligned} \|q(u)\| &\leq (p+1)s^{-p}\|A^{-1}(g)\| \\ &\leq \|A^{-1}\|s^{-p}/(p+1)\|g\| . \end{aligned}$$

Similarly $\|h(u)\| \le p\|A^{-1}\|\|g\|$ and so we have obtained the required estimates on $\|q\|$ and $\|h\|$. Finally the uniqueness of q and h is a trivial consequence of the injectivity of A. □

Remarks.

1. The proof we give here is based on one due to Grauert and Remmert (See R. Narasimhan [3] and also Hörmander [1]). For alternative proofs based on Cauchy's integral formula, we refer the reader to Gunning and Rossi [1], R. Narasimhan [3], Whitney [1] .

2. Notice that if f is any non-identically zero analytic function defined on a neighbourhood of $0 \in \mathbb{C}^n$, we may always find a $\mathbb{C}$-linear change of coordinates on $\mathbb{C}^n$ such that the conditions of the Theorem hold for f. We then say f is *normalised* in direction z_n. Of course, we can always simultaneously normalise any finite set of analytic functions.

3. An immediate consequence of Theorem 3.2.1 is the Weierstrass Division Theorem for germs of analytic functions: Suppose $f \in \mathcal{O}_0$ and that for some representative F of f, $F(0,z_n)$ has a zero of multiplicity p at $z_n = 0$. Then any $g \in \mathcal{O}_0$ may be written uniquely in the form $g = qf + h$, where $q \in \mathcal{O}_0$ and h is the germ of a polynomial in z_n of degree less than p and with coefficients analytic in z'.

Examining the proof of Theorem 3.2.1, we see that the result remains true if we replace $D = D' \times D_s(0)$ by $D'' \times D_s(0)$ where D" is any open neighbourhood of 0 in $\mathbb{C}^{n-1}$ which is contained in D'. Indeed estimates 1 and 3 obviously continue to hold on the smaller neighbourhood $D'' \times D_s(0)$. Moreover, given a finite set $f_1,\ldots,f_k$ of analytic functions defined on a neighbourhood of 0 in $\mathbb{C}^n$, we may, by Remark 2, above, assume that the functions are simultaneously normalised in the z_n-direction. We can then choose a polydisc $D' \times D_s(0)$ in $\mathbb{C}^n$ for which estimates 1, 2 and 3 are valid for all the functions f_j. A straightforward consequence of these observations is the following strengthened form of the Division Theorem that we shall need in §6.

Theorem 3.2.1'. Let $f_1,\ldots,f_k$ be analytic functions defined on a neighbourhood of 0 in $\mathbb{C}^n$ and suppose that each f_j is normalised in direction z_n. Fix a complementary subspace $\mathbb{C}^{n-1}$ of $\mathbb{C}z_n$. Then we may find an open neighbourhood W of 0 in $\mathbb{C}^{n-1}$ and $s > 0$ such that if U is any open neighbourhood of 0 in $\mathbb{C}^{n-1}$ contained in W then the conclusions of the Weierstrass Division theorem hold for all bounded analytic functions on $U \times D_s(0)$ with respect to each divisor f_j, $1 \leq j \leq k$.

Remark. As an exercise the reader may verify that the Division Theorem does not hold for arbitrary neighbourhoods of 0 contained in $W \times D_s(0)$.

Our first application of the Weierstrass Division theorem will be to put the germ of an analytic function in a normalised form. First, we give a definition.

Definition 3.2.2. A germ $P \in \mathcal{O}_0$ is said to be a *Weierstrass polynomial* (of degree p) if (on some neighbourhood of 0)

$$P(z_1,\ldots,z_n) = z_n^p + \sum_{j=0}^{p-1} a_j(z')z_n^j ,$$

where the a_j are analytic functions defined on some neighbourhood of $0 \in \mathbb{C}^{n-1}$ and vanishing at 0 for every j.

We similarly define a Weierstrass polynomial $P \in \mathcal{O}_a$, $a \in \mathbb{C}^n$.

In what follows we let $\mathcal{O}_0'$ denote the ring of germs of analytic functions at $0 \in \mathbb{C}^{n-1}$.

Theorem 3.2.3. (The Weierstrass Preparation Theorem). Let $f \in \mathcal{O}_0$ and suppose that $f(0,z_n)$ has a zero of order $p \geq 0$ at $z_n = 0$. Then there exists a unique unit $h \in \mathcal{O}_0$ and Weierstrass polynomial W of degree p such that $f = hW$. That is, on some neighbourhood of $0 \in \mathbb{C}^n$ we have

$$f(z) = h(z)\left(z_n^p + \sum_{j=0}^{p-1} a_j(z')z_n^j\right)$$

where $h(0) \neq 0$ and $a_j(0) = 0$, $j < p$.

Proof. Take $g = z_n^p$ in Theorem 3.2.1. □

Remarks.

1. There are analogues of the Division and Preparation theorems for differentiatiable functions which were first proved by Malgrange [1].

These results are important in the theory of singularities of differentiable mappings and we refer to Malgrange [1] and Tougeron [1] for further details.

2. In Whitney [2: §10] there is a discussion about the set of directions z_n which lead to a Weierstrass polynomial of minimal degree. In Levinson [1,2] the reader may find results showing that we can make a local holomorphic change of coordinates to put an analytic function in the standard Weierstrass polynomial form.

3. Factorization and finiteness properties of $\mathcal{O}_0$

Let us start by recalling some definitions from algebra.

Definition 3.3.1. Let R be a commutative ring with identity

1. $f \in R$ is said to be *irreducible* or *prime* if any relation $f = gh$, $g,h \in R$, implies that one or other of f and g is a unit.

2. R is said to be a *unique factorization domain* if every $f \in R$ can be written as a finite product of irreducible factors and this decomposition of f is unique up to the order of the factors and multiplication by units.

3. R is said to be *Noetherian* if every ideal $I \triangleleft R$ is finitely generated. That is, if there exist $g_1,\ldots,g_k \in I$ such that

$$I = \{\sum a_j g_j : \text{all } a_j \in R\}.$$

4. If $M \subset R^m$ (R^m denotes the module of m-tuples of elements in R), M is said to be a *submodule* of R^m if M is closed under (coordinate-wise) addition and scalar multiplication by elements of R.

Notation. Given a commutative ring R, R[X] will always denote the ring of polynomials in the indeterminate X and with coefficients in R.

Lemma 3.3.2. Let $W \in \mathcal{O}_0'[z_n]$ be a Weierstrass polynomial and suppose that

$$W = P_1 \ldots P_q ,$$

where $P_j \in \mathcal{O}_0'[z_n]$. Then each P_j is a Weierstrass polynomial up to multiplication by units in $\mathcal{O}_0'$.

Proof. Set $p = \text{degree}(W)$, $p_j = \text{degree}(P_j)$, $1 \le j \le q$. Taking $z' = 0$ we have

$$z_n^p = \prod_{i=1}^{q} P_j(0, z_n) .$$

Since $p = \Sigma p_j$, we must have the coefficient of $z_n^{p_j}$ in P_j non-vanishing at $z' = 0$. Since $\mathbb{C}[z_n]$ is a unique factorization domain, the coefficients of all the lower order terms in P_j vanish at $z' = 0$. Hence $P_j = u_j W_j$, where W_j is a Weierstrass polynomial and u_j, the coefficient of $z_n^{p_j}$ in $P_j(0, z_n)$, is a unit in $\mathcal{O}_0'$. □

Lemma 3.3.3. Let $W \in \mathcal{O}_0'[z_n]$ be a Weierstrass polynomial. Then W is irreducible in the ring $\mathcal{O}_0'[z_n]$ if and only if it is irreducible in $\mathcal{O}_0$.

Proof. The other implication being trivial, it is enough to show that if W is not irreducible in $\mathcal{O}_0$ then it is not irreducible in $\mathcal{O}_0'[z_n]$. Suppose then that $W = fg$, where $f, g \in \mathcal{O}_0$ are not units. By the Weierstrass Preparation theorem we may write $f = UW_1$, $g = VW_2$, where $W_1, W_2 \in \mathcal{O}_0'[z_n]$ are Weierstrass polynomials and U and V are units. Thus $W = uW_1W_2$ for some unit $u \in \mathcal{O}_0$. Since W_1W_2 is a Weierstrass polynomial and $W = uW_1W_2 = 1W$, the uniqueness part of the preparation theorem implies that $u = 1$ and $W = W_1W_2$. Hence we have shown that W is not irreducible in $\mathcal{O}_0'[z_n]$. □

Theorem 3.3.4. $\mathcal{O}_0$ is a unique factorization domain.

Proof. Our proof goes by induction on n. For $n = 0$, $\mathcal{O}_0 = \mathbb{C}$ and the result is trivial. Suppose true for $n-1$. By Gauss' theorem (see Van der Waerden [1; page 70]) $\mathcal{O}_0'[z_n]$ is a unique factorization domain. Let $f \in \mathcal{O}_0$. Then $f = uW$, where u is a unit in $\mathcal{O}_0$ and W is a Weierstrass polynomial. Since $\mathcal{O}_0'[z_n]$ is a unique factorization domain, W may be written as a unique, up to order, product of irreducible polynomials $W_j \in \mathcal{O}_0'[z_n]$: $W = W_1 \dots W_q$. By Lemma 3.3.3, each W_j is irreducible in $\mathcal{O}_0$ and so we have expressed f as a finite product of irreducible elements of $\mathcal{O}_0$. We leave the proof of uniqueness, up to order and multiplication by units, as an exercise for the reader. □

Before proving that $\mathcal{O}_0$ is Noetherian, we recall

Lemma 3.3.5. Let R be a Noetherian ring and M be a submodule of R^p, then M is a finitely generated R-module.

Proof. Let $\pi: R^p \to R$ denote the projection on the first factor and set $M_1 = \pi M \subset R$ and $M_2 = \text{Kernel } \pi|M \subset R^{m-1}$. M_1 is an ideal and so we may find $f_1,\ldots,f_m \in M$ such that the set $\pi(f_1),\ldots,\pi(f_k)$ generates M_1. Let $\tilde{M}_1$ denote the submodule of M generated by $f_1,\ldots,f_k$. The inductive hypothesis implies that M_2 is finitely generated. Since $M = \tilde{M}_1 + M_2$, it therefore follows that M is finitely generated. □

Theorem 3.3.6. The ring $\mathcal{O}_0$ is Noetherian.

Proof. Our proof goes by induction on n. For n = 1, the result is trivially true since every ideal is generated by a power of z_1. Suppose the result is true for $n-1$ and let I be a non-zero ideal of $\mathcal{O}_0$. Changing coordinates if necessary and applying the Preparation theorem we may find a Weierstrass polynomial $W \in I$. For any $f \in I$ we then have by the Division theorem

$$f = q(f)W + r(f) ,$$

where r(f) is a polynomial in z_n of degree less than p, the degree of W. Let $M = \{r(f): f \in I\}$. If we write $r(f) = a_1(z')z_n^{p-1} + \ldots. + a_p(z')$, where $a_j \in \mathcal{O}'_0$, we see that the set of coefficients $(a_1,\ldots,a_p)$ corresponding to polynomials r(f), $f \in I$, defines a submodule of $\mathcal{O}'^p_0$. By Lemma 3.3.5 and our inductive hypothesis, we may find a finite set of generators $(h_{11},\ldots,h_{1p}),\ldots,(h_{s1},\ldots,h_{sp})$ for this submodule. In other words, the set of polynomials

$$p_j(z',z_n) = h_{j1}(z')z_n^{p-1} + \ldots + h_{jp}(z'),\ 1 \le j \le s,$$

generates M over $\mathcal{O}'_0$. Therefore $\{W,p_1,\ldots,p_s\}$ is a finite set of generators for I. □

Exercises.

1. Let U be a connected subset of $\mathbb{C}^n$. Show that if n = 1 or U is a domain of holomorphy then A(U) is not a unique factorization domain. (Hint: For the case n = 1, use the Weierstrass theorem. For domains of holomorphy see the proof of Theorem 2.4.8). In fact A(U) is never a unique factorization domain. For further details on uniqueness of factorization in A(U) we refer the reader to Whitney [1; pages 37-40,332].

2. Let f be a representative of $f_0 \in \mathcal{O}_0$. Show that if $Df(0) \neq 0$, then f_0 is irreducible (Hint: Apply the implicit function theorem).

3. Let $f(y,z) = y^2 - z^2(1-z)$. Show that f_0 is not irreducible even though f is irreducible as a polynomial in $\mathbb{C}[y,z]$.

§4. Meromorphic functions

In this section we apply the theory of §§2,3 to show how we may give a satisfactory definition of a meromorphic function of more than one complex variable.

Lemma 3.4.1. Let $W, P \in \mathcal{O}_0'[z_n]$, $U \in \mathcal{O}_0$ and suppose that $P = WU$. If W is a Weierstrass polynomial then $U \in \mathcal{O}_0'[z_n]$.

Proof. The leading coefficient of W is a unit in $\mathcal{O}_0'$ and so we may apply the polynomial division algorithm to obtain $P = WU' + R$, where $U', R \in \mathcal{O}_0'[z_n]$ and R is of degree less than W. But the uniqueness part of the Weierstrass Division theorem implies that $R = 0$, $U' = U$. □

Proposition 3.4.2. Let f and g be analytic functions defined on some neighbourhood of $0 \in \mathbb{C}^n$. Suppose that f_0 and g_0 are relatively prime in $\mathcal{O}_0$ (that is, they have no common irreducible factor). Then we may find an open neighbourhood D of $0 \in \mathbb{C}^n$ such that

$$f_a, g_a \in \mathcal{O}_a \text{ are relatively prime for all } a \in D.$$

Proof. Without loss of generality we may suppose that f and g are Weierstrass polynomials. By Lemma 3.3.3, f_0 and g_0 are relatively prime in $\mathcal{O}_0'[z_n]$. Since $\mathcal{O}_0'$ is an integral domain, we may form its quotient field M_0'. It follows from Gauss' lemma that f_0 and g_0 are relatively prime in $M_0'[z_n]$. Hence there exist $a_0, b_0 \in \mathcal{O}_0'[z_n]$ and $h_0 \in \mathcal{O}_0'$, $h_0 \neq 0$, such that $h_0 = a_0 f_0 + b_0 g_0$. On some neighbourhood of $0 \in \mathbb{C}^n$ we therefore have

$$h(z') = a(z)f(z) + b(z)g(z) .$$

(As usual, a, b and h denote representatives for the germs a_0, b_0 and h_0 respectively). Since f and g are Weierstrass polynomials we may choose a polydisc neighbourhood D' of $0 \in \mathbb{C}^{n-1}$ such that for all $z' \in D'$ the polynomials $f(z', z_n)$ and $g(z', z_n)$ do not vanish identically. Choose $r > 0$ so that $f, g, a, b, h \in A(D)$, where $D = \{z \in \mathbb{C}^n \colon z' \in D' \text{ and } |z_n| < r\}$. Let $\zeta = (\zeta', \zeta_n) \in D$. Making the substitution $z_n = (z_n - \zeta_n) + \zeta_n$ in a, b, f and g we may consider the relation $h_\zeta = a_\zeta f_\zeta + b_\zeta g_\zeta$ as defining an equation in $\mathcal{O}_0'[z_n - \zeta_n]$. Since f_ζ and g_ζ are not identically zero, we may assume that any common factor of f_ζ and g_ζ is the germ of a Weierstrass polynomial W in $\mathcal{O}_0'[z_n - \zeta_n]$. But then by Lemma 3.4.1, this gives a factorization of h_ζ into a product WP, where $P \in \mathcal{O}_0'[z_n - \zeta_n]$. Hence W must be of degree zero and so equal to 1. Therefore W is a unit. □

Remark. Proposition 3.4.2 gives a beautiful example of a characteristic "coherence" result in complex analysis: An algebraic result that holds at a point tends to hold in a neighbourhood of the point. In other words we have a transition from algebra (results holding in the ring of germs at a point) to a local result (statements holding on an open set).

From now on we shall let M_a denote the quotient field of $\mathcal{O}_a$. Thus $M_a = \{f_a/g_a \colon f_a, g_a \in \mathcal{O}_a \text{ with } g_a \neq 0\}$. Since $\mathcal{O}_a$ is a unique factorization domain each $m \in M_a$ may be written uniquely (up to multiplication by units) in the form f_a/g_a where f_a and g_a are relatively prime.

Suppose that $m \in M_a$. Write $m = f_a/g_a$, where f_a and g_a are relatively prime. If $g_a(a) \neq 0$, we may define the value of m at a, m(a), to be $f_a(a)/g_a(a)$. Clearly m(a) does not depend on our choices for f_a and g_a provided only that they are relatively prime. In case $n = 1$, if $g_a(a) = 0$ then $f_a(a) \neq 0$ and we can define $m(a) = \infty$ (see

Chapter 1, §4, Example 1). For $n > 1$, it is certainly possible for us to have $g_a(a) = f_a(a) = 0$ and then there is no sensible way to define $m(a)$ as the next lemma shows.

Lemma 3.4.3. Let f and g be analytic on some neighbourhood of $0 \in \mathbb{C}^n$, $n > 1$, and suppose $f(0) = g(0) = 0$ and f_0 and g_0 are relatively prime. Then given any $\zeta \in \mathbb{C}$, there exist z arbitrarily close to 0 such that $g(z) \neq 0$ and $f(z)/g(z) = \zeta$.

Proof. Replacing f by $f - \zeta g$, it is no loss of generality to suppose that $\zeta = 0$. We may then assume that f and g are Weierstrass polynomials and, exactly as in the proof of Proposition 3.4.2, we may find $a,b \in \mathcal{O}_0'[z_n]$ and $h \in \mathcal{O}_0'$, $h \neq 0$, such that on some neighbourhood of 0 we have

$$h(z') = a(z)f(z',z_n) + b(z)g(z',z_n) \qquad \ldots.1$$

If the lemma is false, there exists an open neighbourhood D of 0 such that $f(z) = 0$ implies $g(z) = 0$, $z \in D$. But for small z', the polynomial $f(z',z_n)$ certainly has a small zero z_n and so it follows from 1 that $h(z') = 0$ for all z' in some neighbourhood of 0. Contradiction. □

Let us set $M = \bigcup_{a \in \mathbb{C}^n} M_a$.

(Disjoint union over $\mathbb{C}^n$). M is called the *sheaf of germs of meromorphic functions* on $\mathbb{C}^n$.

Definition 3.4.4. Let Ω be an open subset of $\mathbb{C}^n$. A map $m: \Omega \to M$ is said to define a *meromorphic function* on Ω if

1. $m(z) \in M_z$ for all $z \in \Omega$ (m is a "section" of M)

2. For every $z \in \Omega$, there exists $V \in \mathcal{U}_z$ and $f,g \in A(V)$ such that $m(a) = f_a/g_a$, for all $a \in V$.

We denote the set of meromorphic functions on Ω by $M(\Omega)$.

Example 1. Let $\Omega \subset \mathbb{C}^n$ be open and $f,g \in A(\Omega)$, where g is not identically zero on any connected component of Ω. We may define $m \in M(\Omega)$ by

$$m(z) = f_z/g_z, \; z \in \Omega.$$

We often write $m = f/g$.

Proposition 3.4.5. Let $\Omega \subset \mathbb{C}^n$ be open and $m \in M(\Omega)$. We may find an open cover $\{U_j\}$ of Ω and $m_j \in M(U_j)$ such that

1. $m_j = m|U_j$

2. For every j, there exist $f_j, g_j \in A(U_j)$ such that $m_j = f_j/g_j$ and $f_{j,a}$ and $g_{j,a}$ are relatively prime for all $a \in U_j$.

Proof. Immediate from Proposition 3.4.2. □

Remark. Proposition 3.4.5 is, of course, relatively trivial if $n = 1$: Every meromorphic function m on $\Omega \in \mathbb{C}$ may be written in the form $u(z)(z-a)^p$ on some neighbourhood of each point $a \in \Omega$ with $u(a) \neq 0$ and $p \in \mathbb{Z}$. In fact by Corollary 1.3.4 we can write $m = f/g$, with f_a and g_a relatively prime at every point $a \in \Omega$. It is true, though hard to prove, that a meromorphic function on an arbitrary domain in $\mathbb{C}^n$ can be written as a quotient f/g (problem of Poincaré). However, as we shall see later in Chapter 12, it is not true for arbitrary domains in $\mathbb{C}^n$ that we can require f_a and g_a to be everywhere relatively prime.

Notation. If $f_a, g_a \in \mathcal{O}_a$ are relatively prime we shall write $(f_a, g_a) = 1$. More generally, if $f, g \in A(U)$ and $(f_a, g_a) = 1$ for all $a \in U$ we shall write $(f,g) = 1$.

The local description of meromorphic functions given by Proposition 3.4.5 is essentially unique as the next lemma shows.

Lemma 3.4.6. Let $\Omega \subset \mathbb{C}^n$ be open and $m \in M(\Omega)$. Suppose that for some open subset $V \subset \Omega$ we may find $f, f', g, g' \in A(V)$ such that

1. $m|V = f/g = f'/g'$.

2. $(f,g) = 1$, $(f',g') = 1$.

Then there exists $u \in A(V)$ such that

A. $f = uf'$, $g = ug'$.

B. u is non-vanishing on V. That is, u_a is a unit for all $a \in V$.

Proof. At each point $a \in V$ we have $f_a/g_a = f'_a/g'_a$. Hence $f_a g'_a = g_a f'_a$. If f_a is not a unit, f_a must divide f'_a, since $(f_a, g_a) = 1$. Hence there exists a unique $u_a \in \mathcal{O}_a$ such that $u_a f_a = f'_a$. Since f'_a is therefore not a unit and $(f'_a, g'_a) = 1$, there exists $v_a \in \mathcal{O}_a$ such that $v_a f'_a = f_a$. Substituting, we find that $u_a v_a = 1$ and so u_a is a unit. Hence f'_a/f_a is a unit in $\mathcal{O}_a$. Similarly, if g_a is not a unit, g'_a/g_a is a unit in $\mathcal{O}_a$. It follows that for all $a \in V$, there exists a unique unit $u_a \in \mathcal{O}_a$ such that $f'_a/f_a = g'_a/g_a = u_a$. But the equation $f_a u_a = f'_a$ uniquely defines an analytic function u on v satisfying conditions A and B of the lemma. □

Using the above results we are now able to define the pole and zero sets of a meromorphic function and prove that they are analytic subsets (see Definition 2.2.1).

Suppose that $\Omega \subset \mathbb{C}^n$ is open and $m \in M(\Omega)$. By Proposition 3.4.5, we may find an open cover $\{U_j\}$ of Ω and $m_j \in M(U_j)$ such that $m_j = m|U_j$ and $m_j = f_j/g_j$ with $(f_j, g_j) = 1$. We define the subsets $Z(m)$, $P(m)$, $T(m)$ by

$$Z(m) \cap U_j = \{z \in U_j : f_j(z) = 0\}$$

$$P(m) \cap U_j = \{z \in U_j : g_j(z) = 0\}$$

$$T(m) \cap U_j = \{z \in U_j : f_j(z) = g_j(z) = 0\}.$$

We claim that $Z(m)$, $P(m)$ and $T(m)$ are well defined analytic subsets of Ω which depend only on m (not on the local representation of m on U_j). To see this, choose another open cover $\{U'_k\}$ of together with relatively prime $f'_k, g'_k \in A(U'_k)$ satisfying $m|U'_k = f'_k/g'_k$. On the overlap $U_j \cap U'_k$ we have $f_j/g_j = f'_k/g'_k$ and so, by Lemma 3.4.6, there exists a nowhere zero $u \in A(U_j \cap U'_k)$ such that on $U_j \cap U'_k$

$$f_j = uf'_k \text{ and } g_j = ug'_k .$$

But this implies that $f_j(z) = 0$ if and only if $f'_k(z) = 0$ and $g_j(z) = 0$ if and only if $g'_k(z) = 0$. Hence $Z(m)$, $P(m)$ and $T(m)$ are well defined analytic sets.

Definition 3.4.7. Let m be a meromorphic function defined on a domain in $\mathbb{C}^n$. The analytic sets Z(m), P(m) and T(m) constructed above are called the *zero set* of m, *pole set* of m and *indeterminancy set* of m respectively.

Example 2. For n = 1, the indeterminancy set is always empty (see the comments immediately preceding Lemma 3.4.3). For n > 1, the indeterminancy set may be non-empty: Let $f(z_1,z_2) = z_1^2 - z_2^3$, $g(z_1,z_2) = z_1 z_2$. Define $m \in M(\mathbb{C}^2)$ by $m = f/g$. Then $Z(m) = \{(t^3,t^2): t \in \mathbb{C}\}$, $P(m) = \{(z_1,z_2): z_1 = 0 \text{ or } z_2 = 0\}$ and $T(m) = \{(0,0)\}$.

As follows from Lemma 3.4.2 there is no way to define the value of a meromorphic function at points of its indeterminancy set. However, we do have

Proposition 3.4.8. Let m be a meromorphic function on the domain Ω in $\mathbb{C}^n$. Then m defines an analytic function on $\Omega \setminus P(m)$.

Proof. Let $\{U_j\}$ be an open cover of Ω and $f_j, g_j \in A(U_j)$ satisfy the conditions of Proposition 3.4.5. We may define $m: \Omega \setminus P(m) \to \mathbb{C}$ by

$$m(z) = f_j(z)/g_j(z), \quad z \in U_j \setminus P(m) .$$

We leave it to the reader to check that this construction gives a well defined analytic function on $\Omega \setminus P(m)$. □

To conclude this section, we briefly return to the problem of constructing meromorphic functions with specified principal parts or pole and zero sets (see §7 of Chapter 2).

Definition 3.4.9. Let Ω be a domain in $\mathbb{C}^n$ and $\{U_i : i \in I\}$ be an open cover of Ω. Suppose we are given $m_i \in M(U_i)$ such that $m_i - m_j \in A(U_{ij})$ for all $i,j \in I$. The *Cousin I problem* is to construct $m \in M(\Omega)$ such that $m - m_i \in A(U_i)$ for all $i \in I$.

If we can always solve the Cousin I problem on Ω, we call a *Cousin I domain*.

Remark. If Ω is a Cousin A domain, it is certainly a Cousin I domain.

Notation. Let $A^*(U)$ and $M^*(U)$ denote the groups of units in $A(U)$ and $M(U)$ respectively.

Definition 3.4.10. Let Ω be a domain in $\mathbb{C}^n$ and $\{U_i : i \in I\}$ be an open cover of Ω. Suppose we are given $m_i \in M^*(U_i)$ such that $m_i m_j^{-1} \in A^*(U_{ij})$ for all $i,j \in I$. The *Cousin II problem* is to construct $m \in M(\Omega)$ such that $mm_i^{-1} \in A^*(U_i)$ for all $i \in I$.

If we can always solve the Cousin II problem on Ω, we call Ω a *Cousin II domain*.

Remarks.

1. If Ω is a Cousin B domain, it is certainly a Cousin II domain.

2. Suppose that we can construct $m \in M(\Omega)$ satisfying the conditions of the Cousin II problem. Then, with the notation of Definition 3.4.10, we see that

$$Z(m) = \bigcup_{i \in I} Z(m_i)$$

and similarly for the pole and indeterminancy sets. We return in Chapter 5 to the question of multiplicities.

3. The definition of Cousin II domain we have given above is equivalent to that given in Definition 2.7.2. as the reader may easily verify using Proposition 3.4.5.

Exercises.

1. Show that Lemma 3.4.1 need not be true if W is a polynomial in z_n but not a Weierstrass polynomial.

2. Let U be an open subset of $\mathbb{C}^n$ and $f,g \in A(U)$. Suppose that for some $a \in U$, $(f_a, g_a) = 1$. Show that in general $(f,g) \neq 1$.

§5. Local properties of analytic sets

In this section we show how the unique factorization and Noetherian properties of the ring $\mathcal{O}_0$ are reflected in the local structure theory of analytic sets.

We start by examining the case of the zero set of an analytic function f defined on some neighbourhood of 0 in $\mathbb{C}^n$. By uniqueness of factorization, we may write $f_0 = p_1^{r_1} \cdots p_k^{r_k}$, where $p_1, \ldots, p_k \in \mathcal{O}_0$ are distinct primes. Suppose that $f, p_1, \ldots, p_k$ are defined as analytic functions on some neighbourhood U of 0. Since $f = p_1^{r_1} \cdots p_k^{r_k}$ on U we have

$$Z(f) = \bigcup_{j=1}^{k} Z(p_j) .$$

Therefore as a first step in the local study of the zero sets of analytic functions it is natural to examine the zero sets of primes in $\mathcal{O}_0$. Suppose then that $p \in \mathcal{O}_0$ is prime. Changing coordinates if necessary and multiplying by a unit, we may find a polydisc neighbourhood D' of 0 in $\mathbb{C}^{n-1}$ and $s > 0$ such that if $D = \{z \in \mathbb{C}^n\colon z' \in D'$ and $|z_n| < s\}$, then p is a Weierstrass polynomial on D:

$$p(z,z_n) = z_n^p + \sum_{j=0}^{p-1} a_j(z')z_n^j .$$

Let $\delta \in A(D')$ denote the discriminant of p (we recall that δ is the resultant of p and $p' = \partial p/\partial z_n$, δ is a polynomial in the coefficients of p and $\delta(z') = 0$ if and only if $p(z',z_n)$ has a multiple root. For further details we refer the reader to Van der Waerden [1]). Since p is prime, $\delta \not\equiv 0$ (for otherwise p and p' would have a non-constant common factor - see Van der Waerden [1]). Let $\Sigma = \{z' \in D'\colon \delta(z') = 0\}$ (Σ is the *discriminant locus* of p). We let $\pi\colon D \to D'$ denote the projection on the first $n-1$ coordinates. Figure 6 describes $Z(p) \subset D$.

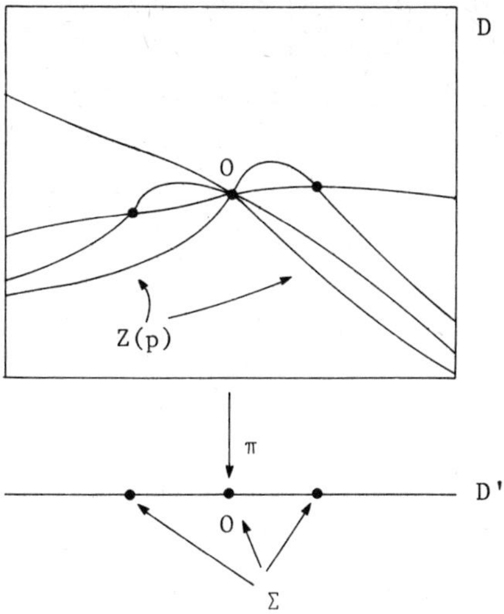

Figure 6.

At each point $z_0' \in D' \setminus \Sigma$, $p(z_0', z_n)$ has precisely p distinct roots. Denote these roots by $\lambda_1(z_0'), \ldots, \lambda_p(z_0')$. Since $\delta(z_0') \neq 0$, $\frac{\partial p}{\partial z_n}(z_0', \lambda_j(z_0')) \neq 0$, $1 \leq j \leq n$, and we may apply the implicit function theorem to deduce that the roots λ_j depend analytically on z', $z' \notin \Sigma$ (see Remark 4, §1 of Chapter 2). It follows that $Z(p) \setminus \pi^{-1}(\Sigma)$ is a (complex) submanifold of D of complex codimension 1 (see Chapter 4 for terminology). Moreover, π clearly restricts to a p-fold covering map of $Z(p) \setminus \pi^{-1}(\Sigma)$ over $D' \setminus \Sigma$. The zero set Z(p) may have "singularities" (points where Z(p) fails to be a submanifold of D) at points lying over Σ.

Examples.

1. Let $f(z,w) = z^2 - w^3$. Then f_0 is prime and is a Weierstrass polynomial in z. The discriminant locus of f is the origin of $\mathbb{C}$ (the w-axis). The reader may verify that at the origin of $\mathbb{C}^2$ it is not possible to find a submanifold chart (analytic or C^∞) for Z(f). However, Z(f) is a topological manifold as the map $t \to (t^3, t^2)$ defines a homeomorphism of $\mathbb{C}$ onto Z(f).

2. Let $f(u,v,w) = u^{2p} - vw$, for some positive integer p. Then f_0 is prime and is a Weierstrass polynomial in u. The discriminant locus

of f is the origin of the (v,w)-plane. We claim that Z(f) is not a topological manifold. If it were, it would have to be modelled on $\mathbb{C}^2$ since $Z(f) \setminus \{0\}$ is of complex dimension 2. Moreover, since $\mathbb{C}^2 \setminus \{0\}$ is simply connected, it would be possible to find simply connected neighbourhoods of 0 in $Z(f) \setminus \{0\}$. But the map $\phi: \mathbb{C}^2 \setminus \{0\} \to Z(f) \setminus \{0\}$ defined by $\phi(t,s) = (ts, t^p, s^p)$ is a p-fold covering map. Hence there is no neighbourhood of zero in $Z(f) \setminus \{0\}$ which is simply connected and so Z(f) cannot be a topological manifold.

We shall now make a more general study of the local structure of analytic sets. Our approach is close to that given in Gunning and Rossi [1]. We start by defining the germ at a point of a subset of $\mathbb{C}^n$. Suppose X,Y are subsets of $\mathbb{C}^n$. We say that X and Y are equivalent at $a \in \mathbb{C}^n$ if there exists $U \in \mathcal{U}_a$ such that $X \cap U = Y \cap U$. This relation is certainly an equivalence relation on subsets of $\mathbb{C}^n$. We denote the equivalence class of a subset X by X_a and refer to X_a as the *germ* of X at a.

If α and β are germs of sets at a, we define

$$\alpha \cup \beta = (X \cup Y)_a; \quad \alpha \cap \beta = (X \cap Y)_a; \quad \alpha - \beta = (X \setminus Y)_a.$$

where X and Y are representatives for α and β respectively. Just as for germs of functions, it is easy to verify that the finite union, intersection and difference of germs of sets are well defined. As an exercise, the reader may show that arbitrary unions or intersections of germs of sets need not be well defined.

We write $\alpha \subset \beta$, it there exist representatives X and Y of α and β respectively such that $X \subset Y$.

Suppose $f_a \in \mathcal{O}_a^p$. Choose a representative $f = (f_1,\ldots,f_p)$ for f_a. We let $Z(f_a)$ denote the germ of Z(f) at a. That is, $Z(f_a) = f^{-1}(0)_a$. $Z(f_a)$ clearly depends only on f_a and not on the choice of representative function for f_a. We let A_a denote the set of all germs at a of analytic sets. Thus $X_a \in A_a$ if and only if there exists $f_a \in \mathcal{O}_a^p$ such that $X_a = Z(f_a)$.

Lemma 3.5.1. Suppose $f = (f_1,\ldots,f_p)$ is a representative function for $f_a \in \mathcal{O}_a^p$. Then

$$Z(f_a) = \bigcap_{i=1}^{p} Z(f_i)_a$$

$$= \bigcap_{i=1}^{p} Z(f_{i,a}).$$ □

(Here $f_{i,a}$ denotes the germ of f_i at a).

Proof. Trivial.

Notation. If $f = (f_1,\ldots,f_p) \in A(U)^p$, we shall adopt the conventions that $Z(f_1,\ldots,f_p) = Z(f)$ and $Z(f_{1,a},\ldots,f_{p,a}) = Z(f_1,\ldots,f_p)_a = Z(f)_a$.

Proposition 3.5.2. Let $X_a, Y_a \in \mathcal{A}_a$. Then $X_a \cup Y_a$, $X_a \cap Y_a \in \mathcal{A}_a$.

Proof. We may suppose that $X_a = Z(f_1,\ldots,f_p)$, $Y_a = Z(g_1,\ldots,g_q)$ for suitable $f_i, g_j \in \mathcal{O}_a$. Clearly

$$X_a \cup Y_a = Z(f_i g_j \colon 1 \le i \le p,\ 1 \le j \le q)$$

$$X_a \cap Y_a = Z(f_1,\ldots,f_p,g_1,\ldots,g_q).$$ □

Defintion 3.5.3. Suppose $X_a \in \mathcal{A}_a$. We say $f_a \in \mathcal{O}_a$ vanishes on X_a if for some representatives X of X_a and f of f_a we have $f|X \equiv 0$. We write $f_a = 0$ on X_a.

Definition 3.5.4. Let $X_a \in \mathcal{A}_a$. We define

$$\mathcal{I}(X_a) = \{f_a \in \mathcal{O}_a \colon f_a = 0 \text{ on } X_a\}.$$

Proposition 3.5.5. Let $X_a \in \mathcal{A}_a$. Then $\mathcal{I}(X_a)$ is an ideal of $\mathcal{O}_a$.

Proof. Trivial. □

Notation. We write $\mathcal{I} \lhd \mathcal{O}_a$ to signify that $\mathcal{I}$ is an ideal of $\mathcal{O}_a$.

Example. Let X be an analytic subset of the domain Ω in $\mathbb{C}^n$. For each $a \in \Omega$, we set $I_a(X) = I(X_a)$. Then $I_a(X) \lhd \mathcal{O}_a$ for every $a \in \Omega$. If $a \notin X$, $I_a(X) = \mathcal{O}_a$.

Before stating the next proposition recall that if R is a commutative ring and $I \lhd R$, then the *radical* of I, Rad(I), is defined by

$$\text{Rad}(I) = \{r \in R\colon \text{For some positive integer } p,\ r^p \in I\}.$$

Proposition 3.5.6. For every germ $X_a \in A_a$

$$I(X_a) = \text{Rad}(I(X_a)).$$

Proof. If f_a^p vanishes on X_a then certainly f_a vanishes on X_a. □

Proposition 3.5.7. Let I be an ideal in $\mathcal{O}_a$ and let $\{a_1,\ldots,a_p\}$ be a finite set of generators for I. Then

1. If $f \in I$, f vanishes on $Z(a_1,\ldots,a_p)$.

2. $Z(a_1,\ldots,a_p)$ depends only on I and not on the particular choice of generators for I.

Proof. Let $f \in I$. Then $f = \sum_{j=1}^{p} f_j a_j$, $f_j \in \mathcal{O}_a$, $1 \le j \le p$. Hence $f \in Z(a_1,\ldots,a_p)$, proving 1. Statement 2 follows straightforwardly from 1. □

Definition 3.5.8. Suppose $I \lhd \mathcal{O}_a$. We define

$$Z(I) = Z(a_1,\ldots,a_p),$$

where $\{a_1,\ldots,a_p\}$ is any finite set of generators for I.

It follows from Propositions 3.5.6 and 3.5.7 that we have a correspondence between radical ideals in $\mathcal{O}_a$ and germs of analytic sets. In the next theorem we list further properties of this correspondence.

Theorem 3.5.9.

1. If $X_a, Y_a \in A_a$ and $X_a \subset Y_a$ then $I(X_a) \supset I(Y_a)$.

1'. If $I, J \lhd O_a$ and $I \subset J$ then $Z(I) \supset Z(J)$.

2. If $X_a, Y_a \in A_a$ and $X_a \neq Y_a$ then $I(X_a) \neq I(Y_a)$.

3. If $X_a \in A_a$, $I(X_a) = \mathrm{Rad}(I(X_a))$.

3'. If $I \lhd O_a$, $Z(I) = Z(\mathrm{Rad}(I))$.

4. If $X_a \in A_a$, then $Z(I(X_a)) = X_a$.

4'. If $I \lhd O_a$, $I(Z(I)) \supseteq \mathrm{Rad}(I)$.

Proof. We shall prove statements 2, 4 and 4'. Statement 3 is Proposition 3.5.6. We leave the remaining cases as exercises for the reader (the full proof may be found in Gunning and Rossi [1]).

Proof of 2. For some $U \in U_a$ and $f_j \in A(U)$, $1 \leq j \leq p+q$, we may write $X = Z(f_1,\ldots,f_p)$, $Y = Z(f_{p+1},\ldots,f_{p+q})$, where X,Y are representatives for X_a, Y_a respectively. Since $X_a \neq Y_a$, we may find $z_W \in ((X \setminus Y) \cup (Y \setminus X)) \cap W$ for every $W \in U_a$. Hence there exists $j(W) \in \{1,\ldots,p+q\}$ such that $f_{j(W)}(z_W) \neq 0$. This is true for all sufficiently small neighbourhoods W of a and so there must exist a j_o such that f_{j_o} does not vanish identically on $((X \setminus Y) \cup (Y \setminus X)) \cap W$ for any $W \in U_a$. If $1 \leq j_o \leq p$, this implies $f_{j_o} \neq 0$ on Y_a and if $p+1 \leq j_o \leq p+q$, $f_{j_o} \neq 0$ on X_a. Hence $I(X_a) \neq I(Y_a)$.

Proof of 4. If $f_a \in I(X_a)$ then certainly $f_a = 0$ on X_a and so $Z(I(X_a)) \supseteq Z_a$. On the other hand we may write $X_a = Z(f_1,\ldots,f_k)$ for some $f_j \in O_a$. Since $f_j = 0$ on X_a, $f_j \in I(X_a))$ and so $Z(f_j) \supseteq Z(I(X_a))$. This holds for all j and so $X_a = Z(f_1,\ldots,f_k) \supseteq Z(I(X_a))$.

Proof of 4'. Obviously, $I(Z(I)) \supseteq I$ for any $I \lhd O_a$. By 3', $Z(I) = Z(\mathrm{Rad}(I))$ and so $I(Z(I)) = I(Z(\mathrm{Rad}(I))) \supseteq \mathrm{Rad}(I)$.

Remark. A consequence of Theorem 3.5.9 is that the map $X_a \to I(X_a)$ is injective *into* the set of radical ideals in O_a with left inverse defined by mapping I to Z(I), for radical $I \lhd O_a$. To show that this map is onto we need to know that $I(Z(I)) = I$ for all radical ideals I of O_a and not just that $I(Z(I)) \supseteq I$ (condition 4'). In fact we do have equality and this fundamental result is the *Nullstellensatz* for germs of analytic sets. The proof of the Nullstellensatz lies much deeper than the other results of this section and may be found in Gunning and Rossi [1], Gunning [1], Hervé [1] or R. Narasimhan [3]. We shall prove a special case, the Nullstellensatz for principal ideals, later in this section and this will suffice for our intended applications.

Definition 3.5.10. We say that a germ $X_a \in \mathcal{A}_a$ is *irreducible* if whenever $X_a = Y_a \cup Z_a$, $Y_a, Z_a \in \mathcal{A}_a$, we have either $X_a = Y_a$ or $X_a = Z_a$.

Theorem 3.5.11. A germ X_a is irreducible if and only if $\mathcal{I}(X_a)$ is prime.

Proof. Suppose $\mathcal{I}(X_a)$ is not prime. Then there exist $f, g \in \mathcal{O}_a$ such that $fg \in \mathcal{I}(X_a)$ but neither f nor g lie in $\mathcal{I}(X_a)$. Now $X_a = X_a \cap Z(fg) = X_a \cap (Z(f) \cup Z(g)) = (X_a \cap Z(f)) \cup (X_a \cap Z(g))$. Since $f, g \notin \mathcal{I}(X_a)$, $X_a \cap Z(f)$ and $X_a \cap Z(g)$ are strictly contained in X_a and so X_a is not irreducible. Conversely, suppose that X_a is not irreducible. Then $X_a = Y_a \cup Z_a$, where Y_a and Z_a are strictly contained in X_a. By 2 of Theorem 3.5.9, $\mathcal{I}(Y_a)$, $\mathcal{I}(Z_a) \neq \mathcal{I}(X_a)$ and so, by 1 of Theorem 3.5.9, $\mathcal{I}(Y_a)$, $\mathcal{I}(Z_a)$ strictly contain $\mathcal{I}(X_a)$. Pick $f \in \mathcal{I}(Y_a) \setminus \mathcal{I}(X_a)$, $g \in \mathcal{I}(Z_a) \setminus \mathcal{I}(X_a)$. Since $fg \in \mathcal{I}(Y_a) \cap \mathcal{I}(Z_a) = \mathcal{I}(X_a)$, we see that $\mathcal{I}(X_a)$ is not prime. □

Theorem 3.5.12. Let $X_a \in \mathcal{A}_a$. We may decompose X_a into a finite union $X_a^1 \cup \dots \cup X_a^k$ of irreducible germs satisfying

1. $X_a^j \not\subset \bigcup_{i \neq j} X_a^i$, for $j = 1, \dots, k$.
2. The X_a^j are uniquely determined (up to order) by X_a.

Proof. Set $I = \mathcal{I}(X_a)$. It follows from the theory of Noetherian rings that we may write $I = \bigcap_{j=1}^{s} I_j$, where the I_j are primary ideals (that is, $\mathrm{rad}(I_j)$ is prime). See, for example, Van der Waerden [2] or Zariski and Samuel [1]. Set $P_j = \mathrm{Rad}(I_j)$. By 4 of Theorem 3.5.9, $X_a = Z(I) = \bigcup_{j=1}^{s} Z(P_j)$. We may throw away any germs $Z(P_j)$ which are contained in the union of the remaining germs to obtain the required decomposition of X_a as a union of irreducible germs. Suppose that

$X_a = X_a^1 \cup \ldots \cup X_a^k = Y_a^1 \cup \ldots \cup Y_a^p$ are two decompositions of X_a as unions of irreducible germs which satisfy condition 1 of the theorem. For $j \in \{1,\ldots,k\}$, $X_a^j = \bigcup_{i=1}^{p} (Y_a^i \cap X_a^j)$ and since X_a^j is irreducible, we must have $X_a^j = Y_a^{\tau(j)} \cap X_a^j$ for some $\tau(j) \in \{1,\ldots,p\}$. Hence $X_a^j \subset Y_a^{\tau(j)}$, $1 \leq j \leq k$. Similarly, $Y_a^i \subset X_a^{\rho(i)}$ for some $\rho(i) \in \{1,\ldots,k\}$, $1 \leq i \leq p$. It follows that for all i,j we have

$$X_a^j \subset Y_a^{\tau(j)} \subset X_a^{\rho\tau(j)} \text{ and } Y_a^i \subset X_a^{\rho(i)} \subset Y_a^{\tau\rho(i)} .$$

Since the decompositions satisfy condition 1, $\rho\tau$ and $\tau\rho$ are the identity maps. Therefore $k = p$ and $X_a^j = Y_a^{\tau(j)}$, $1 \leq j \leq k$. □

We conclude this section by proving three results about germs of analytic sets whose ideals are principal.

Theorem 3.5.13 (Nullstellensatz for principal prime ideals). Let $f_a \in \mathcal{O}_a$ be irreducible. Then $\mathcal{I}(Z(f_a)) = (f_a)$.

Proof. From 4' of Theorem 3.5.9 we already know that $\mathcal{I}(Z(f_a)) \supset (f_a)$. If $g_a \notin (f_a)$, then $(f_a, g_a) = 1$ since f_a is irreducible. Let f and g be representatives for f_a and g_a respectively. By Lemma 3.4.3, we may find z arbitrarily close to a such that $f(z) = 0$ and $g(z) \neq 0$. Hence $g_a \notin \mathcal{I}(Z(f_a))$. Therefore $\mathcal{I}(Z(f_a)) \subset (f_a)$. □

Corollary 3.5.14 (Nullstellensatz for principal ideals). Let $f_a \in \mathcal{O}_a$. Then $\mathcal{I}(Z(f_a) = \mathrm{Rad}((f_a))$.

Proof. Write $f_a = p_1^{r_1} \ldots . p_s^{r_s}$, where the $p_j \in \mathcal{O}_a$ are irreducible and distinct. We certainly have $Z(f_a) = \bigcup_{i=1}^{s} Z(p_i)$ and so, by Theorem 3.15.3, $\mathcal{I}(Z(f_a)) = \bigcap_{i=1}^{s} (p_i) = \mathrm{Rad}(f_a)$. □

Corollary 3.5.15. Let $p_s^{r_1} \ldots p_s^{r_s}$ be the prime factorization of $f_a \in \mathcal{O}_a$. Then $Z(f_a) = \bigcup_{i=1}^{s} Z(p_i)$ is the unique decomposition of $Z(f_a)$ into irreducible germs given by Theorem 3.5.12.

Proof. All we need to check is that for every i, $Z(p_i) \not\subset \bigcup_{j\neq i} Z(p_j)$. By the Nullstellensatz for principal prime ideals, this condition is equivalent to $(p_i) \not\supset (p_1 \cdots p_{i-1}p_{i+1} \cdots p_s)$. This is certainly true since p_i is not divisible by p_j, $i \neq j$. □

Theorem 3.5.16. Let U be a domain in $\mathbb{C}^n$ and suppose that $f \in A(U)$ is not identically zero. Set $X = f^{-1}(0)$. Given $x \in X$, suppose that p_x is a generator of the principal ideal $I_x(X)$. Then we can choose a representative p for p_x, defined on some neighbourhood V of x in U, such that $I_y(X) = (p_y)$ for all $y \in V$.

Proof. It is clearly sufficient to prove the result for the special case when p_x is irreducible. Changing coordinates and multiplying by a unit we may assume that $x = 0$ and that we may find a polydisc neighbourhood D' of 0 in $\mathbb{C}^{n-1}$ and $s > 0$ such that if $D = \{z \in \mathbb{C}^n: z' \in D', |z_n| < s\}$, then p_0 is the germ of a Weierstrass polynomial p on D:

$$p(z',z_n) = z_n^p + \sum_{j=0}^{p-1} a_j(z')z_n^j .$$

Assuming that $D \subset U$, we have $p^{-1}(0) = X \cap D$. The discriminant locus Σ of p is a proper analytic subset of D and so, in particular, $D' \setminus \Sigma$ is an open dense subset of D'. If $y = (y',y_n) \in p^{-1}(0)$ and $y' \notin \Sigma$, then $\partial p/\partial z_n(y',y_n) \neq 0$. Consequently p_y is irreducible (See Exercise 2, §3). On the other hand suppose that $y = (y',y_n) \in p^{-1}(0)$ and $y' \in \Sigma$. Let ϕ_y be a generator of $I_y(X)$. Then for some positive integer k and unit $u_y \in \mathcal{O}_y$, $p_y = u_y\phi_y^k$. This equation holds on a neighbourhood of y in D and so holds for points $z = (z',z_n) \in p^{-1}(0)$ with $z' \notin \Sigma$. Consequently, $k = 1$. Our arguments prove that $I_y(X) = (p_y)$ for all $y \in D$. □

Remark. Theorem 3.5.16 is a special case of the fundamental coherence theorem of H. Cartan which asserts that if $g_x^1,\ldots,g_x^k$ generate $I_x(X)$ then $g_y^1,\ldots,g_y^k$ generate $I_y(X)$ for y in some neighbourhood of x. Here X is an arbitrary analytic subset. The proof of this result is considerably harder than the special case given above and the reader may find proofs in H. Cartan [2], Gunning and Rossi [1], R. Narasimhan [3] and Whitney [1].

Exercises.

1. Suppose $p(z',z_n)$ is an irreducible Weierstrass polynomial of degree p defined on the polydisc $D = D' \times D_s(0)$, with discriminant $\delta \in A(D')$ and discriminant locus Σ. Let $\pi: D \to D'$ denote the projection on D' and set $X = p^{-1}(0)$.

i) Suppose Y is a connected component of $X \setminus \pi^{-1}(\Sigma)$. Show that $\pi|Y$ is an s-fold cover of $D' \setminus \Sigma$ for some s, $1 \leq s \leq p$. (Hint: Fix $z_0' \in D' \setminus \Sigma$ and a root $\lambda_j(z_0') \in Y$. Consider the unique lifts through $\lambda_j(z_0')$ of all closed paths in $D' \setminus \Sigma$ starting at z_0'. Note that $D' \setminus \Sigma$ is connected (Corollary 2.2.3)).

ii) Using the analyticity of the roots of $p(z',z_n)$ together with the Riemann removable singularities theorem, prove that $\bar{Y}$ is the zero locus of a polynomial in z_n with coefficients analytic in $z' \in D'$.

iii) Deduce from ii) that $\bar{Y} = X$ and that $X \setminus \pi^{-1}(\Sigma)$ is connected.

2. Suppose that f(y,z) is a Weierstrass polynomial in z of degree p defined on an open polydisc $D_a(0) \times D_b(0) \subset \mathbb{C}^2$ and that the discriminant locus is equal to $\{0\} \subset D_a(0)$. Set $c = a^{1/p}$. Show that

i) There exists an analytic function $\phi: D_c(0) \to \mathbb{C}$ such that $f(t^p,\phi(t)) = 0$, $t \in D_c(0)$.

ii) If $\omega = \exp(2\pi i/p)$, then given $y = t^p \in D_a(0)$, $f(y,z) = 0$ if and only if $z = \phi(\omega^j t)$ for some integer j, $a \leq j \leq p-1$. (Hint for i): Define $\gamma: D_c(0) \to D_a(0)$ by $\gamma(t) = t^p$. Now show using Q1 and the Riemann removable singularities theorem that lifts to an analytic map $\tilde{\gamma}: D_c(0) \to D_a(0) \times D_b(0)$ with image X). (Observe that if $y = t^p$ and we define $y^{r/p} = t^r$ then we obtain the *fractional power series* or *Puiseaux series* solution $z = \phi(y^{1/p})$ to $f(y,z) = 0$. Further details may be found in Walker [1]).

3. Let $f = (f_1,\ldots,f_p) \in \mathcal{O}_0^p$. Show that if Df(0) is of maximal rank then Z(f) is irreducible.

§6. Modules over $\mathcal{O}_0$.

Suppose that M is a finitely generated $\mathcal{O}_0$-module. Let $U_1^0,\dots,U_{p(0)}^0$ be a set of generators for M and define $U^0 : \mathcal{O}_0^{p(0)} \to M$ by $U^0(f_1,\dots,f_{p(0)}) = \sum_{j=1}^{p(0)} f_j U_j^0$. The sequence $\mathcal{O}_0^{p(0)} \xrightarrow{U^0} M \to 0$ is exact. The kernel of U^0, $\mathrm{Ker}(U^0)$, is a submodule of $\mathcal{O}_0^{p(0)}$ and so, by Lemma 3.3.5, is finitely generated. If $U_1^1,\dots,U_{p(1)}^1$ is a set of generators for $\mathrm{Ker}(U^0)$ we may repeate the above construction to obtain an exact sequence $\mathcal{O}_0^{p(1)} \xrightarrow{U^1} \mathcal{O}_0^{p(0)} \xrightarrow{U^0} M \to 0$. Iterating this construction we obtain a long exact sequence

$$\cdots \xrightarrow{U^{n+1}} \mathcal{O}_0^{p(n)} \xrightarrow{U^n} \cdots \xrightarrow{U^1} \mathcal{O}_0^{p(0)} \xrightarrow{U^0} M \to 0$$

Such a sequence is called a *free resolution* of M (it is also referred to as a *chain of syzygies* for M - see Gunning and Rossi [1] or Nagata [1]). Suppose that for some s, $\mathrm{Ker}(U^{s-1})$ is a *free* $\mathcal{O}_0$-module. That is, $\mathrm{Ker}(U^{s-1}) \cong \mathcal{O}_0^q$ for some integer q. Then we may replace the original free resolution by the finite free resolution

$$0 \to \mathcal{O}_0^q \xrightarrow{\mu} \mathcal{O}_0^{p(s-1)} \longrightarrow \cdots \to \mathcal{O}_0^{p(0)} \to M \to 0,$$

where μ denotes an isomorphism of $\mathcal{O}_0^q$ onto $\mathrm{Ker}(U^{s-1})$. Such a finite free resolution of M is said to have *length* s.

Theorem 3.6.1. (Hilbert Syzygy theorem). Every $\mathcal{O}_0$-module has a free resolution of length n.

Proof. It is sufficient to show that for any free resolution of M,

$$\xrightarrow{U^s} \mathcal{O}_0^{p(s-1)} \xrightarrow{U^{s-1}} \cdots \xrightarrow{U^1} \mathcal{O}_0^{p(0)} \xrightarrow{U^0} M \to 0,$$

$\mathrm{Ker}(U^{n-1})$ is free.

To simplify notation, we set $K_s = \mathrm{Ker}(U^s)$ and let M_s denote the ideal of $\mathcal{O}_0$ generated by $z_1,\dots,z_s$. In particular, $M_n = m_0$, the maximal ideal of $\mathcal{O}_0$.

Step 1. For $s \geq j$, we have $K_s \cap M_j\mathcal{O}_0^{p(s)} = M_jK_s$. Obviously, we have $K_s \cap M_j\mathcal{O}_0^{p(s)} \supset M_jK_s$ and so it remains to verify the reverse inclusion. Our proof goes by induction on j. Suppose $j = 1$, and that $f \in K_s \cap M_1\mathcal{O}_0^{p(s)}$. Then $f = z_1g$, $g \in \mathcal{O}_0^{p(s)}$. Since $U^s(f) = 0$, $z_1U^s(g) = 0$ and so $U^s(g) = 0$. Therefore $g \in K_s$ and $f \in M_1K_s$. Now suppose that we have proved the result for $j-1$ and all $s \geq j-1$. Let $f \in K_s \cap M_j\mathcal{O}_0^{p(s)}$. Then $f = z_1g_1 + \dots + z_jg_j$, $g_1,\dots,g_j \in \mathcal{O}_0^{p(s)}$. Since $f \in K_s$, $U^s(f) = z_1U^s(g_1) + \dots + z_jU^s(g_j) = 0$. Therefore, $z_jU^s(g_j) = -(z_1U^s(g_1) + \dots + z_{j-1}U^s(g_{j-1}))$ and so, dividing through by z_j we see that $U^s(g_j) \in M_{j-1}\mathcal{O}_0^{p(s-1)}$. Now $U^{s-1}U^s = 0$ and so $U^s(g_j) \in K_{s-1} \cap M_{j-1}\mathcal{O}_0^{p(s-1)}$. By the inductive hypothesis, $U^s(g_j) \in M_{j-1}K_{s-1}$. Since $K_{s-1} = \text{Image}(U^s)$, there exist $h_1,\dots,h_{j-1} \in \mathcal{O}_0^{p(s)}$ such that

$$U^s(g_j) = z_1U^s(h_1) + \dots + z_{j-1}U^s(h_{j-1}).$$

Now set $h_j = g_j - z_1h_1 - \dots - z_{j-1}h_{j-1}$. Certainly $h_j \in K_s$. We claim that $f - z_jh_j \in M_{j-1}K_s$. This follows from the inductive hypothesis since $f - z_jh_j = \sum_{i=1}^{j-1} z_i(g_i + z_jh_i)$ and so $f - z_jh_j \in K_s \cap M_{j-1}\mathcal{O}_0^{p(s)} = K_sM_{j-1}$. Since $h_j \in K_s$, $f \in M_jK_s$.

Step 2. Choose a set of generators $U_1,\dots,U_q$ for K_{n-1} which is minimal - that is, no proper subset of $U_1,\dots,U_q$ generates K_{n-1}. Define $U: \mathcal{O}_0^q \to \mathcal{O}_0^{p(n-1)}$ by $U(f_1,\dots,f_q) = \sum_{j=1}^{q} f_jU_j$. We claim that U is an isomorphism onto K_{n-1}. For this it is sufficient to verify that Ker(U) is zero.

Modify the original free resolution of M at the nth. step to obtain the new free resolution

$$\mathcal{O}_0^q \xrightarrow{U} \mathcal{O}_0^{p(n-1)} \xrightarrow{U^{n-1}} \dots \xrightarrow{U^0} M \to 0 \quad .$$

Applying Step 1 to this resolution with $j = n$ we see that

$$\text{Ker}(U) \cap m_0\mathcal{O}_0^q = m_0\text{Ker}(u) \ .$$

In fact, $\text{Ker}(U) \subset m_0\mathcal{O}_0^q$. To see this, suppose $(f_1,\dots,f_q) \in \text{Ker}(U) \subset \mathcal{O}_0^q$. Then

$$0 = U(f_1,\ldots,f_q) = \sum_{j=1}^{q} f_j U_j .$$

Since $U_1,\ldots,U_q$ is a minimal set of generators, no f_j can be a unit. That is, $f_j \in m_0$, $1 \le j \le q$.

We have now shown that $\mathrm{Ker}(U) = m_0\mathrm{Ker}(U)$. This implies that $\mathrm{Ker}(U) = 0$, for otherwise we may find $a_{ij} \in m_0$ such that

$$U_i = \sum_{i=1}^{q} a_{ij} U_j .$$

Therefore $\det[\delta_{ij} - a_{ij}] = 0$ and so, expanding the determinant, we see that $1 \in m_0$ which is a contradiction. □

Remark. The last paragraph of the proof of Theorem 3.6.1 is a special case of

Nakayama's Lemma: Let R be a local ring with maximal ideal m and M be a finite R-mdoule. Then

a) If $M = mM$, $M = 0$.

b) If N is a submodule of M such that $M = N + mM$, $N = M$.

The proof of a) is the same as that used in the last paragraph of the proof of Theorem 3.6.1.Statement b) follows by applying a) to M/N.

Suppose that M is a submodule of $\mathcal{O}_0^p$. Since $\mathcal{O}_0$ is Noetherian, M has a finite set of generators $U_1,\ldots,U_q$ and so we have an exact sequence $\mathcal{O}_0^q \xrightarrow{U} M \to 0$ where U is defined by mapping $(f_1,\ldots,f_q)$ to $\sum_{j=1}^{q} f_j U_j$. The question naturally arises as to whether this sequence is *split* exact. Certainly it is not generally split exact as a sequence of $\mathcal{O}_0$-modules! However, it makes sense to ask whether it splits as a sequence of vector spaces. That is, can we find a $\mathbb{C}$-linear map $\xi\colon M \to \mathcal{O}_0^q$ such that $U\xi$ is the identity map of M? Equivalently, can we write every $m \in M$ in the form $\sum \xi_i(m) U_i$, where ξ_i is linear, $1 \le i \le q$? In view of the continuity conditions in the Weierstrass Division Theorem it is reasonable to ask whether we can topologise the spaces M and $\mathcal{O}_0^q$ and further require that ξ is a *continuous* linear map. In fact we shall prove a result of this type but instead of working with spaces of germs we shall consider modules of bounded analytic functions defined on some neighbourhood of zero in $\mathbb{C}^n$ (for the splitting of sequences of germs and the topologization of $\mathcal{O}_0^p$ and M we refer to Hörmander [1], H. Cartan [2; Seminar 11] and the exercises at the end of this section).

Suppose U is an open neighbourhood of zero in $\mathbb{C}^n$. We let $A_B(U)^p$ denote the $A_B(U)$-module of p-tuples of bounded analytic functions on U. We define a norm on $A^B(U)^p$ by

$$\|f\| = \max_{1\le i\le p} \|f_i\|_U ,$$

where $f = (f_1,\ldots,f_p) \in A_B(U)^p$. Clearly $A_B(U)^p$ is a Banach space in this norm. If M is a submodule of $\mathcal{O}_0^p$, we define

$$M_B(U) = \{f \in A_B(U)^p: f_0 \in M\}$$

$M_B(U)$ is a normed vector subspace of $A_B(U)^p$. However, it is by no means evident that $M_B(U)$ need be closed in $A_B(U)^p$.

Theorem 3.6.2. Let M be a submodule of $\mathcal{O}_0^p$ and W be an open neighbourhood of zero in $\mathbb{C}^n$. Suppose that we are given $U_1,\ldots,U_q \in A(W)^p$ such that the germs of these functions at zero generate M. Then we may find an open neighbourhood $D \subset W$ of zero in $\mathbb{C}^n$ such that the U_j are bounded analytic functions on D such that if we define

$$U: A_B(D)^q \to M_B(D)$$

by $U(f_1,\ldots,f_q) = \sum_{j=1}^{q} f_jU_j$, then the sequence

$$A_B(D)^q \xrightarrow{U} M_B(D) \to 0$$

is split exact as a sequence of normed vector spaces.

Moreover, given any finite set of submodules and generators we can choose an open neighbourhood D of zero that works for all the submodules and generators simultaneously.

Before starting the proof, we remark that the splitting assumption implies that there exists a continuous linear map $\xi = (\xi_1,\ldots,\xi_q): M_B(D) \to A_B(D)^q$ such that every $m \in M_B(D)$ may be written as a sum $\sum_{i=1}^{q} \xi_i(m)U_i$, where $\xi_i: M_B(D) \to A_B(D)$ is continuous linear, $1 \le i \le q$. In particular, there exists $M_i \ge 0$ such that $\|\xi_i(m)\| \le M_i\|m\|$, for all $m \in M_B(D)$, $1 \le i \le q$.

Proof. Our proof goes by induction on n and p. Certainly the theorem is trivially true if $n = 0$. So assume the result for $n-1$ and all positive integers p.

Step 1. The theorem is true for n and $p = 1$. We suppose U_1 is normalised in the z_n-direction and we fix a complimentary subspace $\mathbb{C}^{n-1}$ to $\mathbb{C}z_n$ in $\mathbb{C}^n$. By the Preparation theorem it is no loss of generality to suppose that U_1 is a Weierstrass polynomial in z_n of degree r, say. By the Division theorem for germs (Remark 3, following Theorem 3.2.1), every $f \in M$ may be written uniquely in the form $f = qU_1 + h$, where $q \in \mathcal{O}_0$ and $h \in \mathcal{O}'_0[z_n] \cap M$ is a polynomial of degree $< r$. The set of such germs h defines a $\mathcal{O}'_0$-module M* which we may regard as a submodule of $\mathcal{O}'^r_0$ (see the proof of Theorem 3.3.6). Let $P_1,\ldots,P_s$ be a set of generators for this submodule. We may write

$$P_i = \sum_{j=1}^{q} a_{ij}U_j,$$

for suitable $a_{ij} \in \mathcal{O}_0$, since $M^* \subset M$. Choose an open neighbourhood $W^* \subset W$ of zero such that all the germs P_i, a_{ij} are represented by bounded analytic functions on W*. By Theorem 3.2.1', we can find $s > 0$ and an open neighbourhood V of 0 in $\mathbb{C}^{n-1}$ such that the conditions for the Division theorem hold for the divisor U_1 on any open neighbourhood $D' \times D_s(0)$ of 0 in $\mathbb{C}^n$, provided that $D' \subset V$. By the induction hypothesis applied to M*, we may find an open neighbourhood $D^* \subset V \cap W^*$ of 0 in $\mathbb{C}^{n-1}$ such that if we define $P(f_1,\ldots,f_s) = \sum_{j=1}^{s} f_jP_j$, then the sequence

$$A_B(D^*)^s \xrightarrow{P} M^*_B(D^*) \to 0$$

is split exact. That is, there exists a continuous linear $\xi^*: M^*_B(D^*) \to A_B(D^*)^s$ such that $P\xi^*$ is the identity. We now set $D = D^* \times D_s(0)$. The conditions of the Weierstrass Division theorem hold on D for the divisor U_1. Hence there exist continuous linear maps

$$q,h: M_B(D) \to A_B(D)$$

such that $F = q(F)U_1 + h(F)$, $F \in A_B(D)$. We now define $\xi = (\xi_1,\ldots,\xi_q): M_B(D) \to A_B(D)^q$ by

$$\xi_1(F) = q(F) + \sum_{i=1}^{s} \xi_i^*(h(F))a_{i1}$$

$$\xi_j(F) = \sum_{i=1}^{s} a_{ij}\xi_i^*(h(F))z_n^i .$$

Clearly ξ is a continuous linear map and $U\xi = 1$. In particular, since $U\xi = 1$, U is onto and so we see that the sequence

$$AB(D)^q \underset{\xi}{\overset{U}{\rightleftarrows}} M_B(D) \to 0$$

is split exact. Since Theorem 3.2.1' holds for a finite set of divisors, it is clear that the proof we have given above goes through for any finite set of $\mathcal{O}_0$-submodules and generators. Hence the first inductive step is proved.

Step 2. The theorem is true for n and p > 1. Let $\pi: \mathcal{O}_0^p \to \mathcal{O}_0^{p-1}$ denote the projection on the first $p-1$ coordinates. Set $M' = \pi M$ and $U_j' = \pi U_j$, $1 \le j \le q$. We define $M'' = \{g \in \mathcal{O}_0: (0,\ldots,0,g) \in M\} \cong \ker(\pi|M)$. M' is a submodule of $\mathcal{O}_0^{p-1}$ generated by $U_1',\ldots,U_q'$, and M'' is an ideal of $\mathcal{O}_0$. Let $P_1,\ldots,P_s$ be a set of generators for M''. Since $\mathrm{Ker}(\pi|M) \subset M$, there exist $a_{ij} \in \mathcal{O}_0$ such that

$$(0,\ldots,0,P_i) = \sum_{i=1}^{q} a_{ij}U_j, \; 1 \le i \le s.$$

Choose an open neighbourhood $V \subset W$ of zero such that a_{ij}, P_i are represented by bounded analytic functions in V. By the inductive hypothesis we may find an open neighbourhood $D \subset V$ of zero such that the conclusions of the theorem hold in D for M' and M''. That is, the sequences of normed vector spaces

$$A_B(D)^{p-1} \xrightarrow{U'} M_B'(D) \to 0 \text{ and } A_B(D)^s \xrightarrow{P} M_B''(D) \to 0$$

are split exact. We denote the splitting maps for these sequences by

$$\xi' = (\xi_1',\ldots,\xi_{p-1}'): M_B'(D) \to A_B(D)^{p-1} \text{ and } \xi'' = (\xi_1'',\ldots,\xi_s''): M_B''(D) \to A_B(D)^s.$$

We now define $\xi = (\xi_1,\ldots,\xi_p): M_B(D) \to A_B(D)^p$ by

$$\xi_i(m) = \xi_i'(\pi(m)) + \sum_{j=1}^{s} \xi_j''(m - \pi(m))a_{ji},\ 1 \le i \le q.$$

ξ is obviously a continuous linear map and $U\xi$ = identity. In particular, U is onto and so the sequence

$$A_B(D)^q \underset{\xi}{\overset{U}{\rightleftarrows}} M_B(D) \to 0$$

is split exact.

It is clear from the proof that we may choose D so that the result holds in D for any finite set of $\mathcal{O}_0$-submodules and generators. This proves the second inductive step. □

The following remarkable result is a straightforward consequence of Theorem 3.6.2.

Theorem 3.6.3. Let Ω be a domain in $\mathbb{C}^n$ and F be any subset of $A(\Omega)$. Then $Z(F) = \bigcap_{f \in F} Z(f)$ is an analytic subset of Ω.

Proof. Fiz $z \in \Omega$. Let I_z denote the ideal generated by $\{f_z: f \in F\}$. $\mathcal{O}_z$ is Noetherian and so we may find a finite set $g_1,\ldots,g_p$ of generators for I_z. Since I_z is generated by the germs f_z, $f \in F$, there exists $f_1,\ldots,f_q$ and $h_{ij} \in \mathcal{O}_z$ such that

$$g_i = \sum_{j=1}^{q} h_{ij} f_i,\ i = 1,\ldots,p, \qquad \ldots.1$$

Choosing representatives for g_i and h_{ij} we may suppose that (1) holds on some neighbourhood U of z contained in Ω. By Theorem 3.6.2, we may find an open relatively compact neighbourhood D of z, with $\bar{D} \subset U$, such that for all $f \in F$, there exist bounded analytic functions a_j on D such that $f = \sum_{i=1}^{p} a_i g_i$. But therefore by (1), $f = \sum_{j=1}^{q} a_j' f_j$ on D, where $a_j' = \sum a_i h_{ij}$. It now follows that $Z(f) \cap D \supset Z(f_1,\ldots,f_q) \cap D$. This holds for all $f \in F$ and so $Z(F) \cap D \supset Z(f_1,\ldots,f_q) \cap D$. The reverse inclusion is trivial, since $f_1,\ldots,f_q \in F$. □

Remarks.

1. An alternative proof of Theorem 3.6.3 may be found in Whitney [1; page 100].

2. Theorem 3.6.2 is a special case of a general splitting theorem for coherent sheaves over "privileged" polycyclinders due to Douady. See Douady [1,2].

3. As an example of an important splitting theorem in differential analysis we might mention Mather's splitting theorem for smooth invariants [1]. For more examples and references see Bierstone and Schwarz [1].

Exercises.

1. Show that a finitely generated $\mathcal{O}_0$-module need not be isomorphic to to any submodule of $\mathcal{O}_0^p$, $p \geq 0$ (Hint: $\mathbb{C}$ is a finitely generated $\mathcal{O}_0$-module if we define $f.z = f(0)z$, $f \in \mathcal{O}_0$).

2. Let M be a submodule of $\mathcal{O}_0^p$ with generators $U_1,\dots,U_q$. Show that, with the notation of Theorem 3.5.2, if $A_B(D)^q \xrightarrow{U} M_B(D) \to 0$ is split exact for some open neighbourhood D of zero then $M_B(D)$ is a closed normed subspace of $A_B(D)^p$.

3. (Closure of modules theorem). Suppose M is a submodule of $\mathcal{O}_0^p$ and $F \in A(U)^p$ for some open neighbourhood U of zero in $\mathbb{C}^n$. Show that if F can be uniformly approximated on compact subsets of U by functions whose germs at zero belong to M then $F_0 \in M$ (Hint: Use Q2 and Theorem 3.6.2).

4. Let $f = \sum a_m z^m \in \mathcal{O}_0$. Define $q_m(f) = |a_m|$. Show that

i) q_m defines a semi-norm on $\mathcal{O}_0$, $m \in \mathbb{N}^n$.

ii) $d(f,g) = \sum 2^{-|m|} q_m(f-g)/(1+q_m(f-g))$ defines a metric on $\mathcal{O}_0$.

iii) The completion of $\mathcal{O}_0$ with respect to the metric defined in ii) is the ring $\mathbb{C}[[z_1,\dots,z_n]]$ of formal power series in $z_1,\dots,zn$.

5*. Suppose M is a submodule of $\mathcal{O}_0^p$ and that $U_1,\dots,U_q$ is a set of generators for M. Prove that the sequence $\mathcal{O}_0^q \xrightarrow{U} M \to 0$ is split exact as a sequence of topological vector spaces. Deduce that M is a closed subspace of $\mathcal{O}_0^p$.

CHAPTER 4. COMPLEX MANIFOLDS

Introduction

Much of this chapter is taken up with the discussion of specific examples of complex manifolds. After general definitions given in section 1, we consider complex submanifolds of $\mathbb{C}^n$ in section 2. We prove that the open Euclidean disc and polydisc are biholomorphically inequivalent. Included in this section are also definitions of the classical domains and Stein manifolds. In section 3 we consider projective algebraic manifolds and in section 5 complex manifolds defined as the quotient of a properly discontinuous group action. In section 4 we define complex tori and prove that every one dimensional complex torus is biholomorphic to a cubic curve. In section 6 we develop the structure theory of complex hypersurfaces and in section 7 we define blowing up and give some indication of its use in desingularising analytic sets and in the classification theory of complex manifolds.

§1. Generalities on complex manifolds and analytic sets

In this section we shall review the definitions of complex manifold and submanifold, analytic map and analytic set. For the most part our defintiions are straightforward adaptations of those from differential manifolds and our presentation will therefore be brief.

Throughout this section all topological spaces will be assumed Hausdorff, paracompact and, unless otherwise indicated, connected. If M is a topological space, we let $\mathcal{U}$ denote the topology of M and $\mathcal{U}_a$ denote the subset of $\mathcal{U}$ consisting of all open neighbourhoods of a given point $a \in M$.

Definition 4.1.1. Let M be a topological space. M is said to have the structure of an *n-dimensional complex manifold* if there exists an atlas $A = \{(U_i, \phi_i) : i \in I\}$ of charts on M such that

1. ϕ_i is a homeomorphism of U_i onto the open subset $\phi_i(U_i)$ of $\mathbb{C}^n$ for all $i \in I$.

2. For all $i, j \in I$, $\phi_i\phi_j^{-1}$ is a biholomorphic map of $\phi_j(U_{ij})$ onto $\phi_i(U_{ij})$.

Remarks.

1. We often refer to an n-dimensional complex manifold as being a *complex manifold modelled on* $\mathbb{C}^n$. We also say that the atlas defines a *complex analytic structure* on M. It is always assumed that a complex manifold M comes with a specific complex analytic structure. As we shall see later, a topological space may admit many distinct complex analytic structures.

2. We generally assume that the atlas A is *maximal*. That is, if (U, ϕ) is any chart on M such that $\phi\phi_j^{-1}$ is biholomorphic for all $j \in I$, then $(U, \phi) \in A$.

3. Every n-dimensional complex manifold has the structure of a 2n-dimensional real analytic (therefore, differential) manifold : Forget the complex structure!

Definition 4.1.2. Let M be an m-dimensional complex manifold with atlas A and N be a connected subset of M. We say that N is an *n-dimensional complex submanifold* of M if for every $x \in N$, there exists $(U, \phi) \in A$ such that ϕ maps $U \cap N$ homeomorphically onto an open subset of $\mathbb{C}^n \times \{0\} \subset \mathbb{C}^n \times \mathbb{C}^{m-n} \approx \mathbb{C}^m$.

Remark. If N is an n-dimensional complex submanifold of M then N has the structure of an n-dimensional complex manifold. Indeed, an atlas for N is implicit in the definition.

Example. An open (connected) subset of an n-dimensional complex manifold M is an n-dimensional complex submanifold of M.

Definition 4.1.3. Let M, N be complex manifolds with atlases A, B respectively. A map $f : M \longrightarrow N$ is said to be *holomorphic* or *analytic* if $\zeta f \phi^{-1} : \phi(U \cap f^{-1}(V)) \longrightarrow \zeta(V)$ is analytic for all $(U, \phi) \in A$, $(V, \zeta) \in B$. If f is a homeomorphism and both f and f^{-1} are analytic we say that f is *biholomorphic* and that M and N are *biholomorphic* or *analytically equivalent*.

Notation. Let M be an m-dimensional complex manifold with topology $\mathcal{U}$. If $U \in \mathcal{U}$, we let A(U) denote the ring of $\mathbb{C}$-valued analytic functions on U. Just as in §1, Chapter 3, we let $\mathcal{O}_x$ denote the ring of germs of analytic functions at a point $x \in M$. Using a chart containing x, it is easy to see that $\mathcal{O}_x \cong \mathbb{C}\{z_1,\ldots,z_m\}$ for all $x \in M$ (of course, the isomorphism is no longer canonical). We let $\mathcal{M}_x$ denote the quotient field of $\mathcal{O}_x$ and $\mathcal{M}$ denote the disjoint union of the fields $\mathcal{M}_x$ over M. We call $\mathcal{M}$ the *sheaf of germs of meromorphic functions* on M.

Definition 4.1.4. Let M be a complex manifold. A map $m : M \longrightarrow \mathcal{M}$ is said to define a *meromorphic function* on M if

1. For all $x \in M$, $m(x) \in \mathcal{M}_x$.
2. We may find an open neighbourhood U of every point $x \in M$ and $f, g \in A(U)$ such that $m(y) = f_y/g_y$, $y \in U$.

Notation. We let $\mathcal{M}(M)$ denote the field of meromorphic functions on M. Given $U \in \mathcal{U}$, we let $A^*(U)$, $\mathcal{M}^*(U)$ denote the multiplicative groups of units in $A(U)$, $\mathcal{M}(U)$ respectively.

Let us now indicate some of the main problems in the theory of complex manifolds. The fundamental problem is, of course, the classification of complex manifolds up to biholomorphism. As we shall soon see this is unrealistically hard, even for domains in $\mathbb{C}^n$. Next we might try to find all complex structures, if any, on a given differential manifold. Again, this appears an intractable problem at present (see §§4, 5). Finally, we might ask for necessary conditions on a differential manifold for it to admit a complex structure. Although some important results have been proved, especially about compact 4-manifolds, relatively little is known. For example, it is still unknown whether the six dimensional sphere admits a complex structure though it is known that the only other sphere that can admit a complex structure is the Riemann sphere. We do, however, have one elementary result.

Proposition 4.1.5. If the differential manifold M admits a complex structure then M is even dimensional and orientable.

Proof. M must obviously be even dimensional as $\mathbb{C}^n$ is always of even (real) dimension. As for the orientability, notice that $\mathbb{C}^n$ has a canonical orientation defined by the complex structure J on $\mathbb{C}^n$ (§5, Chapter 1). If M has complex analytic atlas $\{(U_i, \phi_i)\}$, then $D(\phi_i\phi_j^{-1})(\phi_j(x))$ is $\mathbb{C}$-linear for all i, j and so commutes with J. Consequently, M is orientable. Alternatively, the reader may verify that the real determinant of a complex linear map is always positive (see §4, Chapter 5). □

Remark. If a differential manifold is two dimensional then it admits a complex structure if and only if it is orientable. For a proof of this result, depending on the theorem of Korn and Lichtenstein on the existence of isothermal parameters, see Chern [1].

We conclude this section by giving the straightforward generalisation of Definition 2.2.1 to complex manifolds.

Definition 4.1.6. Let Y be a subset of the complex manifold M. We say that Y is an *analytic subset* of M if for each $x \in M$ there exist $U \in \mathcal{U}_x$ and an analytic function $f : U \longrightarrow \mathbb{C}^p$ such that $Y \cap U = f^{-1}(0)$ (p may depend on x).

Remarks.

1. Notice that an analytic subset of M is necessarily closed.

2. In Chapter 6 we define what is meant by an analytic function on an analytic set as well as giving an "intrinsic" definition of an analytic set.

Exercises

1. Show that if M and N are complex manifolds then $M \times N$ has the structure of a complex manifold.

2. Let M be a complex manifold and $m \in \mathcal{M}^*(M)$. Show that the zero, pole and indeterminacy sets of m are well defined analytic subsets of M.

3. Show that the zero set of $y^2 = z^3$ is not a complex submanifold of $\mathbb{C}^2$.

§2. Complex submanifolds of $\mathbb{C}^n$

Every domain in $\mathbb{C}^n$ has the structure of an open complex submanifold of $\mathbb{C}^n$. As we discussed in §4 of Chapter I, every proper simply connected domain of $\mathbb{C}$ is biholomorphic to the open unit disc. We have no such simple classification of simply connected domains in $\mathbb{C}^n$, $n > 1$.

Example 1. Let D denote the open unit polydisc, centre zero, in $\mathbb{C}^n$, $n > 1$, and set $D^* = D \setminus \{(0,\ldots,0,t) : 0 \le t < 1\}$. D and D^* are certainly homeomorphic but cannot be biholomorphic since D is a domain of holomorphy whilst D^* is not (see example 6, §4, Chapter 2).

Even if Ω and Ω' are homeomorphic domains of holomorphy they need not be biholomorphic.

Example 2. (Theorem of Poincaré) Let D and E respectively denote the open unit polydisc, centre zero, and the open unit Euclidean disc, centre zero, in $\mathbb{C}^n$. Then for $n > 1$, D and E are not biholomorphic. Our proof of this inequivalence uses an elementary argument due to Simha [1]. First we remark that if there exists a biholomorphic map of D onto E then there certainly exists a biholomorphic map of D onto E which preserves the origin. This is just a consequence of the existence of biholomorphisms of D taking any prescribed point of D to the origin (That such maps exist is an immediate consequence of their existence for $n = 1$). We shall show that if $f : D \longrightarrow E$ and $g : E \longrightarrow D$ are holomorphic origin preserving maps then $Df(0)(D) \subset E$ and $Dg(0)(E) \subset D$. Assuming this, we see that if f is a biholomorphism with inverse g then we must have $Df(0)(D) = E$ - since $Df(0).Dg(0) = I$. Hence $Df(0)(\partial D) = \partial E$. But this is absurd since ∂D contains linear pieces of positive dimension whilst ∂E does not.

1. $Df(0)(D) \subset E$.

Let $f = (f_1,\ldots,f_n)$, $v \in D$, $u = (u_1,\ldots,u_n) \in E$ and $< \, , >$ denote the standard Hermitian inner product on $\mathbb{C}^n$. Applying Schwarz's lemma to the function $u \longrightarrow \sum u_i f_i(tv)$, we see that $|<\bar{u}, Df(0)(v)>| \leq 1$. This holds for all $u \in E$, therefore $Df(0)(v) \in E$.

2. $Dg(0)(E) \subset D$.

Let $g = (g_1,\ldots,g_n)$ and $u = (u_1,\ldots,u_n) \in E$. Applying Schwarz's lemma to the function $u \longrightarrow f_j(tu_1,\ldots,tu_n)$ we see that $|\sum u_i \partial f_j/\partial z_i(0)| \leq 1$ for $1 \leq j \leq n$. Hence $Df(0)(u) \in D$.

Remarks.

1. It is not hard to show that there exist no proper holomorphic maps of D into E. See Alexander [1] (By *proper* we mean a map such that inverse images of compact sets are compact).

2. For more inequivalence results of the above type and further references see Alexander [1] or R. Narasimhan [2; Chapter 5].

3. In spite of the negative character of the above examples there certainly are some topological restrictions on domains of holomorphy in $\mathbb{C}^n$. For example, if Ω is a domain of holomorphy then $H_r(\Omega, \mathbb{Z}) = 0$, $r > n$, and Ω has the homotopy type of an n-dimensional CW complex (see Andreotti and Frankel [1] and R. Narasimhan [4, 5] for further details). If Ω is a Runge domain (§7, Chapter 2) then $H^r(\Omega, \mathbb{C}) = 0$, $r \geq n$ (Theorem of Serre [1], see also Hormander [1; Theorem 2.7.11]).

Next we turn to the "classical" domains in $\mathbb{C}^n$ which generalise the open unit disc or upper half plane in $\mathbb{C}$. As we shall not need any results from this theory in the remainder of these notes we shall be rather brief and sketchy in our presentation referring the reader to the references for further details and proofs.

Let Ω be a bounded domain in $\mathbb{C}^n$. We let Aut(Ω) denote the group of biholomorphisms of Ω. A fundamental result of H. Cartan [4] is that Aut(Ω) is a locally compact Lie group (see also R. Narashimhan [2]).

Definition 4.2.1. A domain Ω in $\mathbb{C}^n$ is said to be *homogeneous* if Aut(Ω) acts transitively on Ω. That is, given $y, z \in \Omega$, there exists $f \in \text{Aut}(\Omega)$ such that $f(y) = z$. We say Ω is *symmetric* if, in addition, given $z \in \Omega$ there exists $f \in \text{Aut}(\Omega)$ such that $f(z) = z$, $f(y) \neq y$, $y \neq z$, and $f^2 = I$.

E. Cartan [1] classified all the bounded homogeneous domains in $\mathbb{C}^2$ and $\mathbb{C}^3$ proving that any such domain in $\mathbb{C}^2$ is biholomorphic to either the unit polydisc or the unit Euclidean disc and that in $\mathbb{C}^3$ any such domain is biholomorphic to one of the unit polydisc, the unit Euclidean disc, the product of the 2-dimensional unit Euclidean disc with the unit disc, the domain $\text{Im}(z_3) > \sqrt{\text{Im}(z_1)^2 + \text{Im}(z_2)^2}$. These domains are all symmetric. However, it was later shown by Pyatetskii-Shapiro [1] that there exist bounded homogeneous domains in $\mathbb{C}^n$, $n \geq 4$, which are not symmetric (non-countably many for $n \geq 7$). Proofs of these facts may be found in Pyatetskii-Shapiro [1], Siegel [2] and Hua [1].

We say that a bounded symmetric domain is *irreducible* if it cannot be written as a product of symmetric domains of lower dimension. E. Cartan [1] proved that there exist six classes of irreducible symmetric domain. Four of these are referred to as classical since their automorphism groups are classical semi-simple Lie groups. The remaining two classes are exceptional in that one is encountered only in $\mathbb{C}^{16}$, the other only in $\mathbb{C}^{27}$.

We now briefly describe the classical irreducible symmetric domains.

A. Classical domains of the first type. Suppose $n = pq$. Write an element $z \in \mathbb{C}^n$ as a $p \times q$ matrix $Z = [z_{ij}]$. Let I_q denote the identity in $\mathbb{C}^q$ and $\bar{Z}^*$ denote the conjugate transpose of Z. The domain $D_I = \{Z : I_q - \bar{Z}^* Z$ is positive definite$\}$ is a representative classical domain of the first type. Notice that if $p = n$, $q = 1$, we recover the Euclidean unit disc.

B. Classical domains of the second type. Suppose $n = p(p + 1)/2$. In this case we may identify $\mathbb{C}^n$ with the space of $p \times p$ symmetric matrices. The domain $D_{II} = \{Z : I_p - Z\bar{Z}$ is positive definite$\}$ is a representative classical domain of the second type.

C. Classical domains of the third type. Suppose $n = p(p - 1)/2$. In this case we may identify $\mathbb{C}^n$ with the space of $p \times p$ skew-symmetric matrices. The domain $D_{III} = \{Z : I_p + Z\bar{Z}$ is positive definite$\}$ is a representative classical domain of the third type.

D. Classical domains of the fourth type. These exist in all dimensions and a representative domain is given by

$$D_{IV} = \{z \in \mathbb{C}^n : |\sum z_i^2| > \sqrt{2 \sum |z_i|^2 - 1}\}.$$

Finally we describe Siegel domains of the second kind and their relationship with the classical domains (for the general theory we refer to Pyatetskii-Shapiro [1] or Siegel [2]).

Definition 4.2.2. Let $V \subset \mathbb{R}^n$ be an open convex cone which does not contain any real line. We say that a function $F : \mathbb{C}^m \times \mathbb{C}^m \longrightarrow \mathbb{C}^n$ is a *V-Hermitian form* if

1. $F(u, v) = \overline{F(v, u)}$, $u, v \in \mathbb{C}^m$.

2. F is $\mathbb{C}$-linear in the first variable.

3. $F(u, u) = 0$ if and only if $u = 0$.

4. $F(u, u) \in \bar{V}$ for all $u \in \mathbb{C}^m$.

Definition 4.2.3. A Siegel domain of the second kind is a domain in $\mathbb{C}^{n+m}$ consisting of points (y, z) such that $\mathrm{Im}(z) - F(y, y) \in V$ for some fixed V-Hermitian form F on $\mathbb{C}^m \times \mathbb{C}^m$.

Example 3. Let V denote the set of strictly positive real numbers and F the standard Hermitian inner product on $\mathbb{C}^m$. The corresponding Siegel domain of the second kind is the (unbounded) domain $\{(z, t) \in \mathbb{C}^{m+1} : \mathrm{Im}(t) - \langle z, z\rangle > 0\}$. In fact this domain is biholomorphic to a classical domain of the first type, in this case the unit Euclidean disc $\{u \in \mathbb{C}^{m+1} : \sum|u_i^2| < 1\}$. The reader may easily verify that a biholomorphic map between the two domains is given by

$$u_1 = (t - i)/(t + i), \quad u_{j+1} = z_j\sqrt{2}/(t + i), \quad 1 \le j \le m.$$

More generally it may be shown that any Siegel domain of the 2nd. kind is biholomorphic to a bounded domain. A fundamental result due to Pyatetskii-Shapiro [1] states that every bounded homogeneous domain in $\mathbb{C}^n$ is biholomorphic to an (affinely homogeneous) Siegel domain of the second kind.

Siegel domains are of considerable importance in Harmonic analysis, automorphic function theory and the representation theory of semi-simple Lie groups. For further details and references see Pyatetskii-Shapiro [1] and Hua [1]. For an indication of more recent developments, heavily based on several complex variables, see Wells and Wolf [1]. Finally we remark that the kernel functions and automorphism groups of the classical domains are computed in Hua [1].

Next we shall consider closed submanifolds of $\mathbb{C}^n$. First a general result about compact complex manifolds.

Proposition 4.2.4. Let M be a compact complex manifold. Then $A(M) = \mathbb{C}$. That is, every holomorphic function on M is constant.

Proof. Exactly as for the proof of Lemma 1.4.2. □

Corollary 4.2.5. The only compact complex submanifolds of $\mathbb{C}^n$ are isolated points.

Proof. Let M be a compact complex submanifold of $\mathbb{C}^n$. The coordinate functionals $z \longrightarrow z_j$, $1 \leq j \leq n$, restrict to analytic functions on M. Now apply Proposition 4.2.4. to these analytic functions on M. □

A consequence of Corollary 4.2.5 is that the only interesting closed submanifolds of $\mathbb{C}^n$ are non-compact.

Proposition 4.2.6. Let M be a non-compact closed submanifold of $\mathbb{C}^n$. Then

1. Given $x, y \in M$, $x \neq y$, there exists $f \in A(M)$ such that $f(x) \neq f(y)$.

2. M is holomorphically convex : Given a compact subset K of M, $\hat{K} = \{z \in M : |f(z)| \leq \|f\|_K \text{ for all } f \in A(M)\}$ is a compact subset of M.

3. We can define local analytic coordinates on M by means of globally defined analytic functions. That is, given $z \in M$, there exists a neighbourhood U of z in M and $f_1,\ldots,f_m \in A(M)$ such that, when restricted to U, $(f_1,\ldots,f_m) : U \longrightarrow \mathbb{C}^m$ is a complex analytic chart for M.

Proof. 1 and 2 are obvious since $A(M)$ contains the affine $\mathbb{C}$-linear maps on $\mathbb{C}^n$ restriced to M. For 3, just project M onto the tangent space to M at z. □

Definition 4.2.7. A complex manifold which satisfies properties 1, 2 and 3 of Proposition 4.2.6 is called a *Stein manifold*.

Examples.

4. Every domain of holomorphy is a Stein manifold. We may think of Stein manifolds as being a natural generalisation of domains of holomorphy.

5. Every non-compact Riemann surface is Stein by a theorem of Behnke and Stein [1].

Remark. As we shall see later, Stein manifolds have a rich function theory. Roughly speaking, what we can do continuously or differentiably on a Stein manifold we can do complex analytically ("Oka's principle" - see also Chapters 7 and 12).

We have the following basic result analogous to Whitney's embedding theorem.

Theorem 4.2.8. (Bishop [1], R. Narasimhan [6]). Every n-dimensional Stein manifold has a proper analytic embedding in $\mathbb{C}^{2n+1}$. In particular, every Stein manifold is biholomorphic to a closed submanifold of some $\mathbb{C}^n$.

Proof. The proof of this result is considerably harder than that of Whitney's embedding theorem. The main difficulty lies in constructing *proper* maps from a Stein manifold into $\mathbb{C}^m$. For example, if the manifold is 2 complex dimensional there do not exist any proper analytic $\mathbb{C}$-valued maps. We shall not give a proof of the embedding theorem but instead refer the reader to the original references or to Gunning and Rossi [1] and Hörmander [1]. □

Example 6. The open unit disc D in $\mathbb{C}$ is a Stein manifold. We shall construct a proper embedding of D in $\mathbb{C}^3$. The main part of our construction will be to find a proper analytic map $F : D \longrightarrow \mathbb{C}^2$. For $n > 1$, we let D_n denote the open unit disc, centre zero, and radius $1 - 1/n$. We first construct an analytic map $f : D \longrightarrow \mathbb{C}$ such that

$$|f(z)| > n + 1, \quad z \in \partial D_n, \quad n > 1.$$

To do this we shall construct inductively a sequence $f_n \in A(D)$ satisfying

$$\|f_n\|_{D_{n-1}} \leq 2^{-n} \; ; \; \|f_n\|_{\partial D_n} \geq n + 2 + \sum_{j=2}^{n-1} \|f_j\|_{D_n} \quad \ldots\ldots \; (*)$$

Suppose the f_j are constructed for $j < n$. Set $f_n(z) = \left(\frac{(n-1)}{n} b_n z\right)^{p(n)}$. Choose $b_n \in \left(\frac{n}{n-1}, \frac{n}{n-2}\right)$ and $p(n)$ very large so that (*) holds. This completes the inductive step. Now define $f = \sum_{j=1}^{\infty} f_j$. (*) guarantees that f is analytic and that $|f(z)| > n + 1$, $z \in \partial D_n$.

For $n > 1$, set $X_n = \{z \in D_{n+1}\backslash D_n : |f(z)| \leq n + 1\}$ and $Y_n = \{z \in D_n : |f(z)| \leq n + 1\}$. X_n and Y_n are disjoint compact subsets of D and we clearly have $\hat{X}_n = X_n$, $\hat{Y}_n = Y_n$. We now inductively construct a sequence $h_n \in A(D)$ satisfying

$$\|h_n\|_{Y_n} < 2^{-n} \; ; \quad \|h_n\|_{X_n} > 2 + n + \sum_{j=2}^{n-1} \|h_j\|_{X_n} \quad \ldots\ldots (\#)$$

Suppose the h_j are constructed for $j < n$. Define an analytic function q_n on a neighbourhood of $X_n \cup Y_n$ by requiring it to be zero on Y_n and equal to $3 + n + \sum_{j=2}^{n-1} \|h_j\|_{X_n}$ on X_n. The conditions of the Runge Approximation theorem (Theorem A1.8) hold in D for the compact set $X_n \cup Y_n$ and so we can approximate q_n by an analytic function h_n on D such that $\|h_n - q_n\|_{X_n \cup Y_n} \leq 2^{-n}$. Since h_n satisfies the required conditions the inductive step is completed. Define $h = \sum_{j=2}^{\infty} h_j$. Then (#) guarantees that h is an analytic function on D and that

$$|h(z)| > n + 1 \quad \text{on} \quad X_n, \quad n > 1.$$

Define $F = (f, h) : D \longrightarrow \mathbb{C}^2$. We see that

$$|F(z)| = \max\{|f(z)|, |h(z)|\} > n + 1, \quad z \in D_{n+1}\backslash D_n, \; n > 1.$$

Therefore, F is a proper map. To construct the required proper embedding of D in $\mathbb{C}^3$ we define $\phi : D \longrightarrow \mathbb{C}^3$ by $\phi(z) = (z, F(z))$.

Remarks.

1. It can be shown that there exists a proper embedding of D in $\mathbb{C}^2$. Since every proper map is closed, there can be no proper embedding of D in $\mathbb{C}$.

2. The technique used in the example to construct a proper map is actually a special case of the method used to construct proper maps on an arbitary Stein manifold. The reader should note that the difficulty in generalising the example lies in filling up the space with domains to which we can apply an appropriate version of the Runge Approximation theore

Exercises.

1. Construct a proper embedding of the polydisc $D(z;r_1,\ldots,r_n)$ in $\mathbb{C}^{2n+1}$.

2. Show that if we work with real analytic maps then the real Euclidean disc $E(r)$ in $\mathbb{R}^n$ is real analytically diffeomorphic to $\mathbb{R}^n$.

3*. Construct a proper embedding of the Euclidean disc $E(r)$ in $\mathbb{C}^n$ in $\mathbb{C}^{2n+1}$.

4*. Show that the unit disc in $\mathbb{C}$ admits a proper embedding in $\mathbb{C}^2$.

5. (2nd. Riemann removable singularities theorem). Let X be a complex submanifold of the domain Ω in $\mathbb{C}^n$. Show that if $\operatorname{codim}(X) \geq 2$, then every analytic function on $\Omega \backslash X$ extends to Ω(Show that every analytic function on $\Omega \backslash X$ is locally bounded on Ω).

6*. Generalise the result of question 5 to the case when X is an analytic subset of Ω of codimension at least 2 at every point.

§3. Projective algebraic manifolds

The most important examples of compact complex manifolds are given by the complex projective spaces and their closed submanifolds. We start this section by constructing complex n-dimensional projective space.

Let $P^n(\mathbb{C})$ denote the set of complex lines through the origin of $\mathbb{C}^{n+1}$. We have a natural projection map $q : \mathbb{C}^{n+1}\backslash\{0\} \longrightarrow P^n(\mathbb{C})$ defined by: $q(z)$ = Complex line through z and the origin of $\mathbb{C}^{n+1}$. We give $P^n(\mathbb{C})$ the quotient topology. We call $P^n(\mathbb{C})$, with this topology, *complex n-dimensional projective space.*

Proposition 4.3.1. Complex n-dimensional projective space is a compact, connected Hausdorff space, $n \geq 1$.

Proof. Let S^{2n+1} denote the unit sphere in $\mathbb{C}^{n+1}$ relative to the standard Hermitian inner product on $\mathbb{C}^{n+1}$. Every complex line L in $\mathbb{C}^{n+1}$ meets S^{2n+1} in a set homeomorphic to the unit circle S^1. Consequently, $q(S^{2n+1}) = P^n(\mathbb{C})$. Since S^{2n+1} is compact and connected so therefore is $P^n(\mathbb{C})$. We leave the verification that $P^n(\mathbb{C})$ is Hausdorff as an easy exercise. □

Remark. The map $q : S^{2n+1} \longrightarrow P^n(\mathbb{C})$ is called the *Hopf fibration* of S^{2n+1}. In case $n = 1$, it is a well known fact that the map $q : S^3 \longrightarrow S^2$ defines a non-trivial element of $\pi_3(S^2)$ (See, for example, Hirsch [1;page 131]).

Observe that $(z_0,\ldots,z_n)$, $(z_0',\ldots,z_n') \in \mathbb{C}^{n+1}\backslash\{0\}$ define the same point of $P^n(\mathbb{C})$ if and only if there exists $\lambda \in \mathbb{C}^\bullet$ such that $z_i = \lambda z_i'$, $0 \leq i \leq n$. We may think of non-zero $(n+1)$-tuples $(z_0,\ldots,z_n)$ as defining *homogeneous coordinates* on $P^n(\mathbb{C})$. Here it is understood that two non-zero $(n+1)$-tuples define the same point of $P^n(\mathbb{C})$ if and only if they are non-zero complex multiples of one another.

Proposition 4.3.2. Complex n-dimensional projective space has the natural structure of a complex manifold.

Proof. For $0 \leq i \leq n$, define open subsets U_i of $P^n(\mathbb{C})$ by $U_i = \{(z_0,\ldots,z_n) \in P^n(\mathbb{C}) : z_i \neq 0\}$. We define homeomorphisms $\phi_i : U_i \longrightarrow \mathbb{C}^n$ by $\phi_i(z_0,\ldots,z_n) = (z_0/z_i,\ldots,\widehat{z_i/z_i},\ldots,z_n/z_i)$ ($\widehat{\ }$ denotes omission). Clearly $\{U_i : 0 \leq i \leq n\}$ covers $P^n(\mathbb{C})$ and the set $\{(U_i, \phi_i) : 0 \leq i \leq n\}$ is an atlas on $P^n(\mathbb{C})$. A simple computation shows that $\phi_i\phi_j^{-1}$ is biholomorphic (even rational) for all i, j. □

Remark. It is an outstanding problem to determine whether the complex structure on $P^n(\mathbb{C})$ given above is the only one that $P^n(\mathbb{C})$ admits compatible with its standard differential structure. For $n = 1$ this is well known by the Riemann mapping theorem. In higher dimensions only partial results are known. See Hirzebruch and Kodaira [1], Frankel [1], Hartshorne [1], Mori [1], Siu and Yau [1].

Suppose P is a homogeneous polynomial on $\mathbb{C}^{n+1}$. Then $M = \{z \in \mathbb{C}^{n+1} : P(z) = 0\}$ defines a *complex cone* in $\mathbb{C}^{n+1}$. That is, if $z \in M$ so does λz, for all $\lambda \in \mathbb{C}$. Thus, in a natural way, M determines a subset M^* of $P^n(\mathbb{C})$. Any point in M^* corresponds to a line contained in M. Put another way, M^* is just the zero locus of P in homogeneous coordinates.

Definition 4.3.3. Any subset of $P^n(\mathbb{C})$ that may be represented as the common zero locus of a set (not necessarily finite) of homogeneous polynomials defined on $\mathbb{C}^{n+1}$ is called (*projective*) *algebraic* .

We say that a complex submanifold of $P^n(\mathbb{C})$ is *algebraic* if it is an algebraic subset of $P^n(\mathbb{C})$. We say that a complex manifold is *algebraic* if it is biholomorphic to an algebraic submanifold of some complex projective space.

Remark. It follows from Hilbert's basis theorem that every projective algebraic set is the common zero locus of a *finite* set of homogeneous polynomials.

Examples.

1. Let $(a_0,\ldots,a_n) \in P^n(\mathbb{C})$. The zero set of the polynomial $a_0z_0 + \cdots + a_nz_n$ is called a *complex hyperplane*. Every complex hyperplane is biholomorphic to $P^{n-1}(\mathbb{C})$ and the set of complex hyperplanes is in bijective correspondence with $P^n(\mathbb{C})$. Taking the hyperplane $z_0 = 0$, we see that $P^n(\mathbb{C}) = P^{n-1}(\mathbb{C}) \cup U_0$ and iterating this construction we obtain a cell decomposition of $P^n(\mathbb{C})$ from which we can compute the cohomology ring of $P^n(\mathbb{C})$.

2. If $F : \mathbb{C}^{n+1} \longrightarrow \mathbb{C}$ is a homogeneous polynomial then $F(z) = 0$ defines an algebraic submanifold of $P^n(\mathbb{C})$ if (and only if) $DF_z \neq 0$ for all $z \in F^{-1}(0)\setminus\{0\}$. This is a straightforward consequence of the implicit function theorem and the homogeneity of F (Observe that by Euler's theorem, $DF_z(z) = 0$, $z \in F^{-1}(0)$). In particular, the hyperquadric $z_0^2 + \cdots + z_n^2 = 0$ is an algebraic submanifold of $P^n(\mathbb{C})$.

3. The cubic curve $y^2z - 4x^3 + g_2xz^2 + g_3z^3 = 0$ defines an algebraic submanifold of $P^2(\mathbb{C})$ provided that $g_2^3 - 27g_3^2 \neq 0$ (that is, the discriminant of the cubic is non-zero). We shall return to this important class of algebraic curves in §4.

4. Every compact Riemann surface is algebraic. We prove this for surfaces of genus 1 in §4 and in general in Chapter 10 (see also the exercises at the end of §6, Chapter 7). Proofs may also be found in Siegel [1; Chapter 2] and Griffiths and Harris [1; page 213].

Some indication of the significance of projective space may be gauged from the famous theorem of Chow that we shall prove in Chapter 7:

Theorem. Every complex analytic subset of $P^n(\mathbb{C})$ is algebraic.

Remark. We say that a meromorphic function on $P^n(\mathbb{C})$ is *rational* if it can be expressed as the quotient of two homogeneous polynomials of the same degree. One consequence of Chow's theorem is that every meromorphic function on $P^n(\mathbb{C})$ is rational (we have already proved this in §4 of Chapter 1 for the Riemann sphere, $P^1(\mathbb{C})$). We give a proof of the rationality of meromorphic functions on $P^n(\mathbb{C})$, assuming Chow's theorem, in Example 3 of §6.

For the remainder of this section we shall construct a number of important examples of algebraic manifolds. In each case we shall construct a biholomorphic map of the space onto a submanifold of projective space. We shall not verify that the submanifold of projective space is algebraic. This can either be done directly - usually a rather tedious and long computation - or indirectly by citing Chow's theorem.

Examples.

5. For m, n ≥ 1, $P^m(\mathbb{C}) \times P^n(\mathbb{C})$ is algebraic. We take homogeneous coordinates $(z_0,\ldots,z_m)$ and $(w_0,\ldots,w_n)$ on $P^m(\mathbb{C})$ and $P^n(\mathbb{C})$ respectively. We define an embedding ϕ of the product in $P^{nm+n+m}(\mathbb{C})$ by

$$\phi(z_0,\ldots,z_m;w_0,\ldots,w_n) = (z_0w_0,z_0w_1,\ldots,z_0w_n,z_1w_0,\ldots,z_mw_n).$$

We leave it to the reader to verify that ϕ is an embedding (the *Segre* embedding).

6. *Grassmann manifolds.* For $0 \le k \le n$, we let $G_{k,n}(\mathbb{C})$ denote the set of k-dimensional complex linear subspaces of $\mathbb{C}^n$. We call $G_{k,n}(\mathbb{C})$ the *Grassmann manifold* of k-planes in $\mathbb{C}^n$.

We shall first show that $G_{k,n}(\mathbb{C})$ is in bijective correspondence with a closed subset of $P(\wedge^k \mathbb{C}^n)$ (For any complex vector space V, P(V) will denote the projective space of complex lines through the origin of V. Thus $P(\mathbb{C}^n) = P^{n-1}(\mathbb{C})$). Suppose $E \in G_{k,n}(\mathbb{C})$ and let $v_1,\ldots,v_k$ and $w_1,\ldots,w_k$ be bases for E. Then $v_1 \wedge\ldots\wedge v_k = aw_1 \wedge\ldots\wedge w_k$ for some $a \in \mathbb{C}^*$. Conversely, if such a relation holds, $v_1,\ldots,v_k$ and $w_1,\ldots,w_k$ span the same k-dimensional linear subspace of $\mathbb{C}^n$. Consequently we may define an injection $\Theta : G_{k,n}(\mathbb{C}) \longrightarrow P(\wedge^k \mathbb{C}^n)$ by setting $\Theta(E) = v_1 \wedge \dots \wedge v_k$, where $v_1, \dots, v_k$ is any basis for E. $\Theta(G_{k,n}(\mathbb{C}))$ is obviously a closed subset of $P(\wedge^k \mathbb{C}^n)$ and we give $G_{k,n}(\mathbb{C})$ the induced topology.

If we let $e_1, \dots, e_n$ denote the standard basis for $\mathbb{C}^n$ and choose a basis $v_1, \dots, v_k$ for E we can write

$$\Theta(E) = \sum a_{i_1 \dots i_k} e_{i_1} \wedge \dots \wedge e_{i_k}$$

where the $a_{i_1 \dots i_k}$ are skew-symmetric in the indices and are defined uniquely by E up to multiplication by elements of $\mathbb{C}^\bullet$. The scalars $a_{i_1 \dots i_k}$ are called the *Cayley-Plücker-Grassmann* coordinates on $G_{k,n}(\mathbb{C})$.

Next we shall construct a complex analytic atlas for $G_{k,n}(\mathbb{C})$. We let $\underline{k}$ denote the set of (ordered) subsets of $\{1, \dots, n\}$ containing precisely k elements. If $\alpha = \{\alpha_1, \dots, \alpha_k\} \in \underline{k}$, we set $c\alpha = \{1, \dots, n\} \backslash \alpha$ and let E_α denote the k-dimensional subspace of $\mathbb{C}^n$ spanned by the coordinate vectors $e_{\alpha_1}, \dots, e_{\alpha_k}$. Let V_α denote the vector space of $\mathbb{C}$-linear maps from E_α to $E_{c\alpha}$. Clearly $V_\alpha \cong \mathbb{C}^{k(n-k)}$. We have a natural homeomorphism ψ_α of V_α onto an open subset U_α of $G_{k,n}(\mathbb{C})$ defined by mapping a linear function to its graph. Set $\phi_\alpha = \psi_\alpha^{-1}$. Then it is not difficult to show that $\{(U_\alpha, \phi_\alpha) : \alpha \in \underline{k}\}$ is a complex analytic atlas on $G_{k,n}(\mathbb{C})$. Relative to the complex structure defined by this atlas it is now a straightforward matter to verify that the bijection Θ defined above is a complex analytic embedding of $G_{k,n}(\mathbb{C})$ onto a compact complex submanifold of $P(\wedge^k \mathbb{C}^n)$.

We shall now give two other descriptions of $G_{k,n}(\mathbb{C})$ in terms of Lie groups. Observe that $GL(n, \mathbb{C})$ acts transitively on $G_{k,n}(\mathbb{C})$. Given $E \in G_{k,n}(\mathbb{C})$ we let $I(E)$ denote the isotropy subgroup of $GL(n, \mathbb{C})$ at E (that is, $I(E) = \{A \in GL(n, \mathbb{C}) : A(E) = E\}$). It follows from the elementary theory of homogeneous spaces (or directly, in this case) that $G_{k,n}(\mathbb{C}) = GL(n, \mathbb{C})/I(E)$ (Kobayashi and Nomizu [2]). Let $GL(k, n\text{-}k, \mathbb{C})$ denote the subgroup of $GL(n, \mathbb{C})$ consisting of all matrices of the form $\begin{pmatrix} A & B \\ O & C \end{pmatrix}$ where $A \in GL(k, \mathbb{C})$, $C \in GL(n\text{-}k, \mathbb{C})$ and B is any $k \times n$-matrix. If we take E to be the k-plane defined by the first k coordinate vectors, it is easy to see that $I(E) = GL(k, n-k, \mathbb{C})$. Hence,

$$G_{k,n}(\mathbb{C}) \cong GL(n, \mathbb{C})/GL(k, n-k, \mathbb{C})$$

Consequently, $G_{k,n}(\mathbb{C})$ is a complex manifold of dimension $n^2 - k^2 - (n-k)^2 - k(n-k) = k(n-k)$.

Similarly, taking the standard Hermitian inner product on $\mathbb{C}^n$, we see that

$$G_{k,n}(\mathbb{C}) \cong U(n)/(U(k) \times U(n-k))$$

($U(p)$ denotes the unitary group of $\mathbb{C}^p$). Since the unitary groups are compact, it follows from this representation that $G_{k,n}(\mathbb{C})$ is compact.

Grassmann manifolds are particularly important as they are classifying spaces for vector bundles. See the exercises at the end of the section.

The homology of a Grassmann manifold is described using the "Schubert Calculus". See Griffiths and Harris [1] and also Chern [1].

7. *Flag manifolds*. Let $0 < p_1 < p_2 < \cdots < p_k \leq n$ be a sequence of integers. The *flag manifold* $F(p_1,\cdots,p_k,n)$ is defined to be the set of all nested sequences $L_{p_1} \subset L_{p_2} \subset \cdots \subset L_{p_k} \subset \mathbb{C}^n$ of linear subspaces of $\mathbb{C}^n$ with dimensions given by the subscripts. Clearly $F(p, n) = G_{p,n}(\mathbb{C})$. $F(p_1,\cdots,p_k,n)$ has the structure of a compact complex manifold of dimension $\sum_{j=1}^{k-1} p_j(p_{j+1} - p_j)$. The easiest way to prove this is to represent the flag manifold as a homogeneous space as we did for the Grassmann manifold. Flag manifolds are also algebraic. The proof of this is similar to that we gave for the Grassmann manifold. Flag manifolds are important in the representation theory of semi-simple Lie groups. See, for example, Wells and Wolf [1].

Exercises.

1. Let $z_0 \in P^n(\mathbb{C})$. Show that if $n > 1$, $P^n(\mathbb{C})\backslash\{z_0\}$ is an example of a non-compact complex manifold with no non-constant analytic functions.

2. Let $U = \{(z, L) \in \mathbb{C}^n \times G_{k,n}(\mathbb{C}) : z \in L\}$ and define $\pi : U \longrightarrow G_{k,n}(\mathbb{C})$ by $(z, L) = L$. Show that U has the natural structure of a k-dimensional holomorphic vector bundle over $G_{k,n}(\mathbb{C})$. We call U the *universal bundle* or *canonical bundle* of $G_{k,n}(\mathbb{C})$. It can be shown that given any k-dimensional vector bundle E over a differential manifold X there exists a positive integer N, depending only on X, such that for some differentiable map $f : X \longrightarrow G_{k,N}(\mathbb{C})$ we have $f^*U \cong E$. Moreover, f^*U depends, up to isomorphism, only on the homotopy class of the map $f : X \longrightarrow G_{k,N}(\mathbb{C})$. Similar results hold for continuous vector bundles over X. See Atiyah [1], Husmoller [1] and Milnor [1] for further details and proofs.

§4. Complex tori

Let $\omega = \{\omega_1, \ldots, \omega_{2n}\}$ be a real basis of $\mathbb{C}^n$ and let L_ω denote the lattice subgroup of $\mathbb{C}^n$ defined by

$$L_\omega = \left\{ \sum_{j=1}^{2n} m_j\omega_j : (m_1, \ldots, m_{2n}) \in \mathbb{Z}^{2n} \right\}.$$

Clearly $L_\omega \cong \mathbb{Z}^{2n}$. We define the *complex n-torus* T_ω to be the quotient space $T_\omega = \mathbb{C}^n/L_\omega$. The quotient map $\pi : \mathbb{C}^n \longrightarrow T_\omega$ is a local homeomorphism and T_ω is a compact Hausdorff space. The projection π induces a complex structure on T_ω with respect to which π is holomorphic, in fact locally biholomorphic (see also Theorem 4.5.2). Clearly T_ω is diffeomorphic to the standard real 2n-dimensional torus $T^{2n} = \mathbb{R}^{2n}/\mathbb{Z}^{2n}$. Since L_ω is a subgroup of $\mathbb{C}^n$, T_ω inherits a group structure from that on $\mathbb{C}^n$. The group operations on T_ω are obviously analytic and so T_ω has the structure of a *complex Lie group* (that is, a Lie group which is a complex manifold and whose group operations are analytic).

Example 1. Every compact connected complex Lie group is a complex torus and is, in particular, Abelian : By standard results in Lie group theory it is enough to prove that a compact connected complex Lie group G is Abelian. To see that this is so consider, for $g \in G$, the map $\phi_g : G \longrightarrow G$ defined by $\phi_g(h) = ghg^{-1}h^{-1}$, $h \in G$. Now $\phi_e(h) = e$ for all $h \in G$ and so there exists a neighbourhood U of e and a coordinate chart (V, ζ) for G containing e such that $\phi_g(G) \subset V$ for all $g \in U$. But then, by the argument of Proposition 4.2.4 applied to $\zeta\phi_g$, we see that

ϕ_g must be constant on G for all $g \in U$. Hence, by analytic continuation, ϕ_g is constant on G. Therefore, $\phi_g(h) = e$ for all $g, h \in G$ and so G is Abelian.

Proposition 4.4.1. Let $\{\omega_1, \ldots, \omega_{2n}\}$, $\{\omega'_1, \ldots, \omega'_{2n}\}$ be real bases of $\mathbb{C}^n$ and L, L' and T, T' respectively denote the corresponding lattices and tori. Then T is biholomorphic to T' if and only if there exists $A \in GL(n, \mathbb{C})$ such that $A(L) = L'$.

Proof. Suppose $f : T \longrightarrow T'$ is biholomorphic. It is no loss of generality to suppose that f maps the identity of T to the identity of T' (otherwise replace f by $f(e)^{-1}f$). Let $F : \mathbb{C}^n \longrightarrow \mathbb{C}^n$ be any lifting (necessarily biholomorphic) of f to the universal cover $\mathbb{C}^n$ of T, T' :

$$\begin{array}{ccc} \mathbb{C}^n & \xrightarrow{\quad F \quad} & \mathbb{C}^n \\ \Big\downarrow \pi & & \Big\downarrow \pi' \\ T & \xrightarrow{\quad f \quad} & T' \end{array}$$

Clearly $F(z + \omega_j) - F(z) \in L'$, $1 \le j \le 2n$, for all $z \in \mathbb{C}^n$. By continuity, these differences are all constant functions. Hence, for $z \in \mathbb{C}^n$, we have

$$\frac{\partial F}{\partial z_i}(z + \omega_j) = \frac{\partial F}{\partial z_i}(z), \quad 1 \le i \le n \ ; \ 1 \le j \le 2n.$$

By this periodicity, $\partial F/\partial z_i$ is a bounded holomorphic map and so, by Liouville's theorem, constant. Hence

$$F(z) = Ax + b,$$

for some $A \in GL(n, \mathbb{C})$, $b \in \mathbb{C}^n$. Since $f(e) = e'$, $F(0) \in L'$ and so $b \in L'$. We may suppose $b = 0$ (otherwise replace F by $F - b$). By our construction, $A(L) = L'$. The converse is trivial. □

Remark. Notice that the proof above shows that two complex tori are biholomorphic if and only if they are isomorphic as complex Lie groups.

We shall now examine complex tori of dimension one in somewhat greater detail.

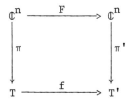

Let $\omega = \{\omega_1, \omega_2\}$ be a real basis for $\mathbb{C}$ and set $\underline{\tau} = \{\omega_2/\omega_1, 1\} = \{\tau, 1\}$. Certainly $\mathrm{Im}(\tau) \neq 0$ and by Proposition 4.4.1, T_ω is biholomorphic to $T_{\underline{\tau}}$. Since $\{\tau, 1\}$ and $\{-\tau, 1\}$ generate the same lattice, we may without loss of generality suppose that $\mathrm{Im}(\tau) > 0$. Hence the set of complex one-dimensional tori is parametrized by points in the upper half plane $H = \{\tau : \mathrm{Im}(\tau) > 0\}$. We shall now determine when two points in the upper half plane determine the same torus. Let $\tau, \tau' \in H$ and L, L', T, T' denote the corresponding lattices and tori. T is biholomorphic to T' if and only if there exists $a \in GL(1, \mathbb{C}) \approx \mathbb{C}^{\bullet}$ such that $a(L) = L'$. That is, if and only if there exist $a_{11},\ldots,a_{22} \in \mathbb{Z}$ such that

$$a\tau = a_{11}\tau' + a_{12}$$

$$a = a_{21}\tau' + a_{22}$$

$$a_{11}a_{22} - a_{21}a_{12} = 1.$$

Therefore τ and τ' define biholomorphic tori if and only if

$$\tau = \frac{a_{11}\tau' + a_{12}}{a_{21}\tau' + a_{22}} \quad \ldots\ldots\ldots\ldots\ldots\ldots\ldots\ldots * $$

We see that the relation (*) defines an action of $SL(2, \mathbb{Z})$ on H. The orbit space of this action, $H/SL(2, \mathbb{Z})$, gives an effective parametrization of the set of complex structures on the 2-dimensional real torus. We say that a subset $D \subset H$ is a *fundamental region* for the action of $SL(2, \mathbb{Z})$ on H if no two points in D lie on the same $SL(2, \mathbb{Z})$ orbit and every $SL(2, \mathbb{Z})$ orbit meets D. It is not hard to show that the region D in the figure below is a fundamental region for $SL(2, \mathbb{Z})$ acting on H (for details see Lang [2] or Robert [1]).

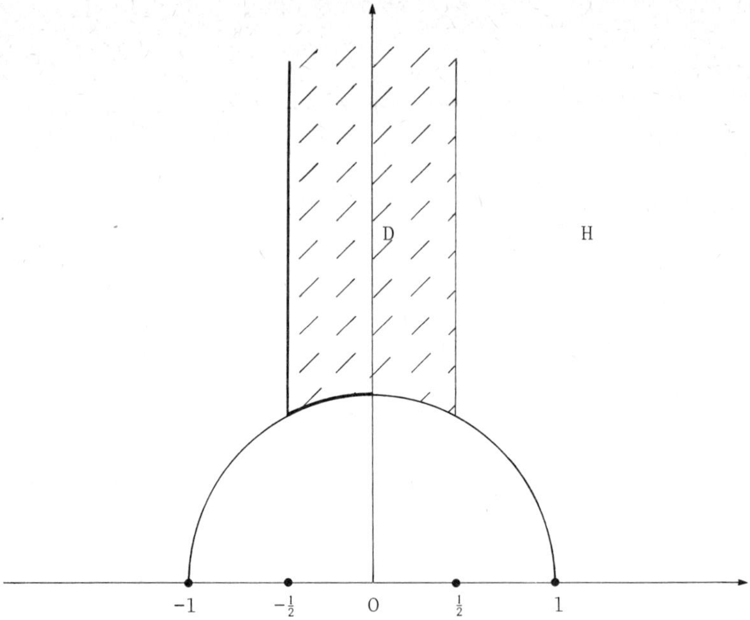

Figure 7

$D = \{\tau \in H : \text{Either} -\frac{1}{2} \leq \text{Re}(\tau) < \frac{1}{2} \text{ and } |\tau| > 1 \text{ or } |\tau| = 1 \text{ and } \text{Re}(\tau) \leq 0\}.$

It is clear from what we have said above that we can "continuously deform" complex structures on a real 2-dimensional torus to obtain an uncountable family of distinct complex structures. This is very characteristic : There may exist many different complex structures on a given differential manifold. The classification of all such structures is generally a problem of great difficutly and only partial results are known. For example, if M is a compact Riemann surface of genus $g \geq 2$, then it is known by the work of Riemann, Teichmüller [1], Rauch [1], Ahlfors [2] and Bers [1] that the set of complex structures on M has the natural structure of a complex analytic space of dimension (3g-3). This space of complex structures - The *moduli space* - will have singularities. Unfortunately, no such global existence theorem holds in general for higher dimensional compact manifolds. One approach to a global theory is due to Griffiths [1]. There is an infinitessimal theory of deformations of complex structure due to Kodaira, Spencer and Kuranishi. This is described in the book by Kodaira and Morrow [1]. Finally we remark the difficult theorem of Douady's [1] to the effect that the set of all complex analytic subspaces of a given complex analytic space has the natural structure of a complex analytic space.

We shall now briefly consider the question of the existence of meromorphic functions on a complex torus. First we look at 1-dimensional tori.

Let $\{\omega_1, \omega_2\}$ be a real basis of $\mathbb{C}$ and let L denote the corresponding lattice. Set $T = \mathbb{C}/L$ and let $\pi : \mathbb{C} \longrightarrow T$ denote the quotient map. Given $a \in \mathbb{C}$, we let P_a denote the parallelogram with vertices a, $a + \omega_1$, $a + \omega_2$, $a + \omega_1 + \omega_2$. P_a is called a *period parallelogram* (for L) and obviously $\pi(P_a) = T$.

Suppose $m \in M(T)$. Clearly $\tilde{m} = m\pi \in M(\mathbb{C})$ and $\tilde{m}$ is L-invariant. That is, given $z \in \mathbb{C}$,

$$\tilde{m}(z + \omega) = m(z), \quad \text{for all} \quad \omega \in L.$$

Equivalently, $\tilde{m}$ is *doubly periodic* with periods ω_1, ω_2,

$$m(z + \omega_1) = m(z + \omega_2) = m(z), \text{ all } \quad z \in \mathbb{C}.$$

We shall say that an L-invariant meromorphic function on $\mathbb{C}$ or, equivalently, a meromorphic function on T, is (L-) *elliptic*. Clearly the set of poles and zeroes of an elliptic function is finite (mod L).

Theorem 4.4.2. Let $m \in M^*(T)$ and $\operatorname{div}(m) = \sum_{i=1}^{k} n_i \cdot z_i$. Then

1. $\sum_{i=1}^{k} \operatorname{residue}_{z_i}(m) = 0.$

2. $\deg(\operatorname{div}(m)) = 0.$

3. $\sum_{i=1}^{k} n_i z_i = e$ (relative to the group law on T).

Proof. Choose $a \in \mathbb{C}$ such that the boundary of the period parallelogram P_a is disjoint from the poles and zeroes of m. 1 follows by integrating m (strictly, $\tilde{m} = \pi m$) round ∂P_a and observing that the integrals along opposite sides cancel by the periodicity of m. Similarly 2 follows by integrating the elliptic function m'/m round ∂P_a (see also Lemma 1.4.2). For 3 we integrate zm'/m round ∂P_a. In this case the integrals round opposite sides do not cancel but we do have

$$(2\pi i)^{-1}\left(\int_a^{a+\omega_1} - \int_{a+\omega_2}^{a+\omega_1+\omega_2} zm'/m\,dz\right) = -(2\pi i)^{-1}\omega_2 \int_a^{a+\omega_1} m'/m\,dz$$

Now $(2\pi i)^{-1}\int_a^{a+\omega_1} m'/m\,dz$ is just the winding number of the closed curve $m([a, a+\omega_1])$ about zero and is therefore an integer, say m_1. Similarly for the other pair of boundary integrals. Lifting to P_a and choosing $\tilde{z}_i \in P_a$ such that $\pi(\tilde{z}_i) = z_i$, $1 \le i \le k$, we see that

$$\sum_{i=1}^{k} n_i \tilde{z}_i = m_2\omega_1 - m_1\omega_2 \in L . \qquad \square$$

Remarks. It follows from condition 1 of the Theorem that an elliptic function cannot have a single pole of order 1. The simplest possibilities are a single double pole with zero residue or two single poles with opposite residue. From condition 2 it follows that if $m \in M^*(T)$, $c \in \mathbb{C}$, then m and m-c have the same number of zeroes. Hence the map $m : T \longrightarrow P^1(\mathbb{C})$ (Chapter 1, §4, Example 1) takes all complex values the same number of times. Condition 3 gives a necessary condition on a divisor in order that it be the divisor of an elliptic function. It turns out that conditions 2 and 3 of the theorem are sufficient conditions on a divisor for it to be the divisor of a meromorphic function (Abel's Theorem).

We shall now construct some elliptic functions. We start by defining *Weierstrass' elliptic* function $\wp(z)$:

$$\wp(z) = z^{-2} + \sum{}' \left((z-\omega)^{-2} - \omega^{-2}\right), \quad z \in \mathbb{C}\setminus L \quad .$$

Here the sum is over all $\omega \in L' = L\setminus\{0\}$. This sum is absolutely convergent at all points of $\mathbb{C}\setminus L$ and converges uniformly on compact subsets of $\mathbb{C}\setminus L$ (See the exercises at the end of this section and note that terms in the sum are $O(\omega^{-3})$). Hence $\wp \in M(\mathbb{C})$. Clearly $\wp$ is even : $\wp(-z) = \wp(z)$, $z \in \mathbb{C}\setminus L$. To show that $\wp$ is elliptic we consider the derivative $\wp'$. Now

$$\wp'(z) = -2\sum (z-\omega)^{-3}$$

is obviously elliptic and so, by integrating, we see that

$$\wp(z+\omega) = \wp(z) + C(\omega),$$

where $C(\omega)$ is independent of z. Since $\wp$ is even, $z = -\omega/2$ gives $C(\omega) = 0$ and so $\wp$ is elliptic. Set

$$G_j = \sum{}' \omega^{-2j}, \quad j \geq 2 .$$

The reader may easily verify that the Laurent expansions of $\wp$ and $\wp'$ at zero are given by

$$\wp(z) = z^{-2} \sum_{j\geq 1} (2j+1)G_{2j+2}z^{2j}$$

$$\wp'(z) = -2z^{-3} + \sum_{j\geq 1} (2j+1)2jG_{2j+2}z^{2j-1}$$

and converge for $0 < |z| < \min_{L'}|\omega|$.

Let us find the divisor of $\wp'$. Since $\wp'$ is odd and L-invariant $\wp'(z) = 0$ if $-z \equiv z \bmod L$. That is, if $2z \in L$. Thus representative zeroes of $\wp'$ in the period parallelogram P_0 are given by $\omega_1/2$, $\omega_2/2$, $(\omega_1+\omega_2)/2$. These are the only zeroes in P_0 by Theorem 4.4.2. Now $\wp$ has only two zeroes in P_0 and, by condition 3 of Theorem 4.4.2 they must be $\{a, -a\}$ for some $a \in T$. These zeroes are not known explicitly in terms of the lattice L.

Proposition 4.4.3. The elliptic functions $\wp$ and $\wp'$ are related by the differential equation

$$\wp'^2 = 4\wp^3 - g_2\wp - g_3,$$

where $g_2^3 - 27g_3^2 \neq 0$ and $g_2 = 60G_4$, $g_3 = 140G_6$.

Proof. $\wp'^2 - 4\wp^3$ is certainly elliptic. Using the Laurent series for $\wp$ and $\wp'$ we easily compute that

$$\wp'^2 - 4\wp^3 = -60G_4z^{-2} - 140G_6 + O(z^2)$$

$$= -g_2\wp - g_3 + h,$$

where h is holomorphic, elliptic and $O(z^2)$. Therefore $h \equiv 0$.

Now we know the zeroes of $\wp'$ and so those of $4\wp^3 - g_2\wp - g_3$. But the zeroes of $\wp'$ are distinct and so therefore are those of the cubic $4x^3 - g_2x - g_3 = 0$. Therefore the discriminant of the cubic, $-16(g_2^3-27g_3^2)$, is non-zero. □

We have already indicated in the previous section that the cubic curve $y^2z - 4x^3 + g_2xz^2 + g_3z^3 = 0$ is non-singular in $P^2(\mathbb{C})$ provided that $g_2^3 - 27g_3^2 \neq 0$. The next theorem gives an explicit embedding of the torus T as an algebraic submanifold of $P^2(\mathbb{C})$.

Theorem 4.4.4. The mapping $\phi : T \longrightarrow P^2(\mathbb{C})$ defined by

$$\phi(z) = (z^3\wp(z),\ z^3\wp'(z),\ z^3)$$

is a biholomorphic embedding of T onto the cubic curve $y^2z - 4x^3 + g_2xz^2 + g_3z^3 = 0$.

Proof. By Proposition 4.4.3, ϕ maps T into the given cubic curve. Certainly ϕ is onto since $\wp(z) = A$ has a solution for every $A \in P^1(\mathbb{C})$ (remarks following Theorem 4.4.2). ϕ is holomorphic since $z^3\wp(z)$ and $z^3\wp'(z)$ are holomorphic and the derivative of ϕ is everywhere of maximal rank as is seen by noting that $\wp''(z) \neq 0$ if $z \in \{\omega_1/2, \omega_2/2, (\omega_1+\omega_2)/2\}$ as these points are simple zeroes of $\wp'$. It remains to be proved that ϕ is injective. Suppose $\wp(z) = \wp(y)$. If $z \not\equiv -z$, mod L, and $y \not\equiv z$, mod L, we see that $y \equiv -z$, mod L, since $\wp$ is even and takes each value twice (counting multiplicities). But $\wp'$ is odd and so $\wp'(z) \neq \wp'(y)$ unless $\wp'(z) = 0$ which cannot happen as $z \not\equiv -z$, mod L. Now suppose $z \equiv -z$, mod L. Since $\wp$ takes each value twice, it is easy to see that $y \equiv -y$, mod L . But if $z \equiv -z$, mod L, $z \in \{\omega_1/2, \omega_2/2, (\omega_1+\omega_2)/2\}$. Now $\wp(\omega_1/2)$, $\wp(\omega_2/2)$, $\wp(\omega_1 + \omega_2/2)$ are the roots of the cubic $4x^3 - g_2x - g_3 = 0$ and are distinct by our assumption on the discriminant. Hence $z \equiv y$, mod L. Our argument proves that ϕ is injective. □

Remarks.

1. Here we have just given a tiny fragment of the theory of elliptic curves. The reader may consult Lang [2], Robert [1] or Siegel [1] for more complete treatments of elliptic curves and functions. Whittaker and Watson [1] is a classical reference.

2. Although we have not proved it here, it is not difficult to show that every meromorphic function on a complex torus is a rational function of $\wp$ and $\wp'$. In particular, any two non-zero meromorphic functions are algebraically dependent (see also Chapter 5, §9).

3. It can be shown that any two meromorphic functions on a compact Riemann surface are algebraically dependent (see Gunning [1; Theorem 26, §10]). Consequently, we may generalise the method of proof of Theorem 4.4.4 to construct a holomorphic map of a compact Riemann surface M onto an algebraic curve in $P^2(\mathbb{C})$ provided that there exist "sufficiently many" meromorphic functions on M. The existence of an abundance of meromorphic functions on M is a consequence of the Riemann-Roch theorem which we prove in Chapter 10 (see also §5, Chapter 7). In general, we cannot embed M onto a non-singular curve in $P^2(\mathbb{C})$ (we always can embedd in $P^3(\mathbb{C})$). We shall show in Chapter 7, §5, that every compact Riemann surface can be represented as a branched cover of $P^1(\mathbb{C})$. In case the surface has genus $g > 1$, and admits a 2-fold branched cover over $P^1(\mathbb{C})$, the surface is called *hyperelliptic*. We have more to say about these matters in Chapter 10. See also Griffiths and Harris [1], Hartshorne [2] and Gunning [1].

For the remainder of this section we shall briefly consider complex tori of dimension > 1.

We no longer have a good moduli space for n-dimensional complex tori if $n > 1$. In fact the natural parameter space for complex structures on a real 2n-dimensional torus turns out to be non-Hausdorff - see Kodaira and Spencer [1; pages 408-414] and also Kodaira and Morrow [1; pages 22-23].

Definition 4.4.5. A complex torus which is algebraic is called an *Abelian variety*.

Every 1-dimensional complex torus is an Abelian variety (Theorem 4.4.4). However, it is not the case that every n-dimensional complex torus is algebraic for $n > 1$; in fact "most" are not. An obvious necessary condition for a complex torus to be algebraic is that it admit non-constant meromorphic functions (of course, we might expect to use such functions to construct maps into projective space). A stronger necessary condition is that we can separate points on the torus using meromorphic functions (since we can on projective space).

Example 2. (Siegel [2]). Let T be the complex 2-dimensional torus whose lattice is generated by $\{(1,0),(0,1),i(\sqrt{2},\sqrt{3}),i(\sqrt{5},\sqrt{7})\} \subset \mathbb{C}^2$. Then T admits no nonconstant meromorphic functions. In particular, T is not algebraic. We shall not prove this result here but instead refer the reader to Siegel [2]. See also Cornalba [1; page 85], de La Harpe [1, page 142], Griffiths and Harris [1] and Swinnerton-Dyer [1].

In fact in every dimension greater than 1 there exist complex tori with no non-constant meromorphic functions. We shall now give necessary and sufficient conditions for a complex torus T with lattice L generated by $\omega_1,\ldots,\omega_{2n}$ to be an Abelian variety.

Let $A : \mathbb{C}^n \times \mathbb{C}^n \longrightarrow \mathbb{R}$ be a real skew-symmetric bilinear form (that is, A is real linear in each variable separately and $A(x, y) = -A(y, x)$. We say that A is a *Riemann form* for T (or L) if
1. $A(L, L) \subset \mathbb{Z}$; 2. $A(ix, y)$ is a symmetric positive definite form on $\mathbb{C}^n$.

Example 3. Every lattice $L \subset \mathbb{C}$ admits a Riemann form. Indeed, suppose that L is generated by $\{\omega_1, \omega_2\}$ and that $\mathrm{Im}(\omega_1/\omega_2) > 0$. Denote the area of the period parallelogram defined by ω_1, ω_2 by S. Clearly $S = \mathrm{Im}(\omega_1\bar{\omega}_2)$. If we define $A : \mathbb{C} \times \mathbb{C} \longrightarrow \mathbb{R}$ by $A(y, z) = S^{-1}\mathrm{Im}(y\bar{z})$, then A is a Riemann form for L. As an exercise the reader may verify that A is the only Riemann form for L which takes the values ± 1 on any basis of L.

For $n > 1$, lattices do not generally admit a Riemann form or even a non-zero positive semi-definite form taking integer values on the lattice. However, it may be shown that the set of all lattices which admit a non-zero Riemann form is dense in the set of all lattices (obvious topology). If A is a Riemann form for L then we have a positive definite Hermitian form H on $\mathbb{C}^n$ whose imaginary part is integer valued on the lattice. Indeed, we may define

$$H(y, z) = A(iy, z) + iA(y, z), \quad y, z \in \mathbb{C}^n .$$

We have the fundamental result

Theorem. A complex torus is algebraic if and only if it admits a Riemann form.

We shall prove that the existence of a Riemann form implies that the torus is algebraic in §6 of Chapter 7 (the converse is not difficult). The theorem will also follow from our results on compact Kähler manifolds presented in Chapter 10. The reader may also find proofs in Cornalba [1], Griffiths and Harris [1], Swinnerton-Dyer [1] and Weil [1]. These references also give further information and references about complex tori. In Chapter 5 we shall have a little to say about the construction of meromorphic functions on complex tori as well as the rôle played by theta functions and holomorphic line bundles.

Exercises.

1. Show that if L is a lattice in $\mathbb{C}$ then $\sum_{L\setminus\{0\}} |\omega|^{-\lambda}$ converges provided that $\lambda > 2$. Deduce that the series for $\wp(z)$ converges uniformly on compact subsets of $\mathbb{C}\setminus L$.

2. Let L be a lattice in $\mathbb{C}$ and choose $\omega \in L\setminus\{0\}$ as close as possible to the origin. Show that there is a basis for L which contains ω.

3. Show that every even elliptic function may be written uniquely as a product $c\prod_{j=1}^{k} (\wp(u_i) - \wp(z))^{p_j}$, $c \in \mathbb{C}$ (You will need part 2 of Theorem 4.4.2.). By writing an elliptic function as the sum of an even elliptic function plus $\wp'$ × even elliptic function, show that every elliptic function is a rational combination of $\wp$, $\wp'$ and that the field of elliptic functions is a quadratic extension of $\mathbb{C}(\wp)$ (Use Proposition 4.4.3).

4. Let F(x, y, z) be homogeneous of degree d and $C \subset P^2(\mathbb{C})$ denote the corresponding curve of degree d. Suppose $X \subset P^2(\mathbb{C})$ is an elliptic curve parametrized by $(\wp(u), \wp'(u), 1)$ as in Theorem 4.4.4. Prove that the number of points of intersection of X with C (counting multiplicities) is 3d (This is a special case of Bezout's theorem.
Hint : Suppose $F(0, 1, 0) \neq 0$. Consider the elliptic function $F(\wp, \wp', 1)$ and apply Theorem 4.4.2, part 2).

5. Suppose that the elliptic curve $X \subset P^2(\mathbb{C})$ is parametrized by $(\wp, \wp', 1)$. Show that the points $u, v, w \in X$ are collinear if and only if $u + v + w = e$ (relative to the group law on X). Deduce that

$$(u + v) = - (u) - (v) + \frac{1}{4}\left(\frac{'(u) - '(v)}{(u) - (v)}\right)^2$$

provided that $u \not\equiv \pm v \pmod{L}$, where L is the lattice associated to X.

6. Use the methods of questions 4 and 5 to study the intersection of conics with a fixed elliptic curve in $P^2(\mathbb{C})$.

7. Let the lattice $L \subset \mathbb{C}$ have basis $\{1, \tau\}$, $\mathrm{Im}(\tau) > 0$. Show that the set of continuous homomorphisms of the torus $\mathbb{C}/L$ is in bijective correspondence with the set $M(L)$ of complex numbers α such that $\alpha L \subset L$. Clearly $M(L) \supset \mathbb{Z}$. Say that $\mathbb{C}/L$ has *complex multiplication* if $M(L)$ is strictly bigger than $\mathbb{Z}$. Show that if $\mathbb{C}/L$ has complex multiplication then $\tau \in \mathbb{Q}(\sqrt{-p})$, for some positive integer p and that $M(L)$ is then a subring of the ring of integers of the field $\mathbb{Q}(\sqrt{-p})$. Deduce that there are only countably many complex tori that admit complex multiplication.

§5. Properly discontinuous actions.

Let M be a complex manifold and Aut(M) denote the group of biholomorphic transformations of M. Denote the identity of Aut(M) by I. Suppose that Γ is a subgroup of Aut(M). We say that Γ acts *freely* on M if the only element of Γ that has fixed points is the identity. That is, if $g \in \Gamma$ and $g(x) = x$ for some $x \in M$ then $g = I$. We say that Γ acts *properly discontinuously* on M if for any pair K_1, K_2 of compact subsets of M, the set $\{g \in \Gamma : g(K_1) \cap K_2 \neq \emptyset\}$ is finite.

Lemma 4.5.1. Let Γ act freely, properly discontinuously on M. Then every point $x \in M$ has an open neighbourhood U such that if $g \in \Gamma$ and $g(U) \cap U \neq \emptyset$, then $g = I$.

Proof. Let V be any relatively compact open neighbourhood of x. Since Γ acts properly discontinuously, $P = \{g \in \Gamma : g(V) \cap V \neq \emptyset\}$ is finite. Define

$$U = V \setminus \bigcup_{g \in P\setminus\{I\}} g^{-1}\overline{(g(V) \cap V)} .$$

We leave it to the reader to check that U satisfies the conditions of the lemma. □

Theorem 4.5.2. Let M be a complex manifold and Γ be a subgroup of $\mathrm{Aut}(M)$ which acts freely, properly discontinuously on M. Then there exists a natural complex structure on M/Γ such that the orbit map $\pi : M \longrightarrow M/\Gamma$ is locally biholomorphic.

Proof. Let $y = \pi(x) \in M/\Gamma$. Choose a chart (U, ϕ) for M containing x such that the conditions of Lemma 4.5.1 hold for the open set U. U is mapped homeomorphically onto $\pi(U)$ by π. Clearly $(\pi(U), \phi^{-1}(\pi|U)^{-1})$ is a chart on M/Γ containing y. Repeating this construction for every point $y \in M/\Gamma$ we obtain a complex analytic atlas on M/Γ relative to which π is locally biholomorphic. □

Examples.

1. Let M be a Riemann surface (not necessarily compact). Then, by the Uniformization theorem, the universal covering surface $\widetilde{M}$ of M is biholomorphic to either the Riemann sphere, upper half plane (or open unit disc) or complex plane. Set $\Gamma = \pi_1(M)$. Then Γ acts freely, properly discontinuously on $\widetilde{M}$ and M is biholomorphic to $\widetilde{M}/\Gamma$. We remark that with the exception of the Riemann sphere, punctured complex plane, complex plane and torus, $\widetilde{M}$ is biholomorphic to the upper half plane. In this case Γ is a discrete subgroup of the group of biholomorphic transformations of the upper half plane, $PSL(2, \mathbb{R})$. ($PSL(2, \mathbb{R}) = SL(2, \mathbb{R})/\{I, -I\}$ and acts on the upper half-plane by mapping $z \longrightarrow (az + b)/(cz + d)$, $ad - bc = 1$). Viewed in this light, the classification of Riemann surfaces amounts to the classification of all discrete subgroups of $PSL(2, \mathbb{R})$.

2. Let Ω be a bounded domain in $\mathbb{C}^n$ and suppose that $\Gamma \subset \mathrm{Aut}(\Omega)$ acts freely, properly discontinuously on Ω and Ω/Γ is compact. Then Ω/Γ is algebraic. We prove this fundamental result of Kodaira [1] in Chapter 10. We remark that it implies that if the universal cover of a compact complex manifold M is biholomorphic to a bounded domain in $\mathbb{C}^n$ then M is algebraic.

3. If $L \in \mathbb{C}^n$ is a lattice subgroup ($\cong \mathbb{Z}^{2n}$), then L acts freely, properly discontinuously on $\mathbb{C}^n$ and so $\mathbb{C}^n/L$ has the structure of a complex manifold; see §4 of this chapter.

4. Let $a_0,\ldots,a_n \in \mathbb{C}^\bullet$ and suppose $|a_j| < 1$, $0 \le j \le n$. Let Γ denote the cyclic subgroup of $\mathrm{Aut}(\mathbb{C}^{n+1}\setminus\{0\})$ generated by $g : (z_0,\ldots,z_n) \longrightarrow (a_0z_0,\ldots,a_nz_n)$. Then Γ acts freely, properly discontinuously on $\mathbb{C}^{n+1}\setminus\{0\}$ and so $\mathbb{C}^{n+1}\setminus\{0\}/\Gamma$ has the structure of a complex manifold which is easily seen to be compact and diffeomorphic to $S^{2n+1} \times S^1$. Any complex manifold diffeomorphic to $S^{2n+1} \times S^1$ is called a *Hopf manifold* and, if $n = 1$, a *Hopf surface*. As we shall see in Chapter 10, Hopf manifolds are never algebraic ($n \neq 0$). Hopf surfaces have been studied in great detail by Kodaira [2]. See also Kodaira and Spencer [1] and Ueno [1 ; pages 235-238]. Here we shall describe the important (and unpleasant!) phenomenon of "jumping" of complex structure that can occur on complex manifolds of dimension greater than one. Let D denote the open unit disc in $\mathbb{C}$ and choose $a \in \mathbb{C}^\bullet$, $|a| < 1$. For $t \in D$, let Γ_t be the subgroup of $\mathrm{Aut}(\mathbb{C}^2\setminus\{0\})$ generated by $g_t:(z_0,z_1) \longrightarrow (az_0+z,az_1)$ Then Γ_t acts freely, properly discontinuously on $\mathbb{C}^2\setminus\{0\}$ and so $M_t=\mathbb{C}^2\setminus\{0\}/\Gamma_t$ has the structure of a complex manifold which is easily seen to be a Hopf surface. In this case, we may define $M = (\mathbb{C}^2 \times \{0\}) \times D/\Gamma$, where Γ is generated by $g : ((z_0, z_1), t) \longrightarrow ((az_0 + tz_1, az_1), t)$. M has the structure of a complex manifold and if we let $\pi : M \longrightarrow D$ denote the natural projection we see that π is holomorphic and $\pi^{-1}(t) = M_t$, $t \in D$. Thus we have defined a complex analytic family of Hopf surfaces, parametrized by points in the unit disc. Suppose $f : M_t \longrightarrow M_s$ is biholomorphic, $t, s \in D$. Lift f to the universal cover to obtain a biholomorphic map $F : \mathbb{C}^2\setminus\{0\} \longrightarrow \mathbb{C}^2\setminus\{0\}$ satisfying $Fg_t = g_sF$. By Hartog's theorem (Theorem 2.3.2), F extends to an analytic map $\bar{F} : \mathbb{C}^2 \longrightarrow \mathbb{C}^2$ which is easily seen to be biholomorphic and to preserve the origin. Again we have $\bar{F}g_t = g_s\bar{F}$. Differentiating this relation at zero we see that $D\bar{F}_0\cdot g_t = g_s\cdot D\bar{F}_0$. Setting $[D\bar{F}_0] = [F_{ij}]$, we easily compute that this relation implies that

$$tF_{00}z_1 = sF_{10}z_0 + sF_{11}z_1$$

$$tF_{10}z_1 = 0 \quad .$$

Hence if $st \neq 0$, we must have $F_{10} = 0$ and $tF_{00} = sF_{11}$. If these relations are satisfied, we see immediately that DF_0 induces a biholomorphic map between M_t and M_s. That is, we have shown that the Hopf surfaces M_t are all biholomorphic provided $t \neq 0$. However, if say $s = 0$, $t \neq 0$, the conditions above imply that $F_{00} = F_{10} = 0$ and so F cannot be biholomorphic. Therefore, M_0 is not biholomorphic to M_t, $t \neq 0$. In other words the complex structure on M_t "jumps" as t passes through zero. Notice that although the analytic map $\pi : M \longrightarrow D$ is differentiably trivial it is *not* analytically trivial.

5. *Calabi-Eckmann manifolds* (Calabi and Eckmann [1]). Fix positive integers p, $q \geq 0$ and consider the projection $\pi : S^{2p+1} \times S^{2q+1} \rightarrow P^p(\mathbb{C}) \times P^q(\mathbb{C})$ obtained as the product of the Hopf fibrations of S^{2p+1} and S^{2q+1}. For each $x \in P^p(\mathbb{C}) \times P^q(\mathbb{C})$, the fibre $\pi^{-1}(x)$ is diffeomorphic to a real two-dimensional torus and so admits a complex structure. Since the base carries a complex structure we might hope to be able to define a complex structure on the product $S^{2p+1} \times S^{2q+1}$ and we shall now show that this is possible.

Fix $\tau \in \mathbb{C}$, $\mathrm{Im}(\tau) > 0$ and let T_τ denote the torus with lattice generated by $\{1, \tau\}$. Given integers r, s, $0 \leq r \leq p$, $0 \leq s \leq q$, define

$$V_{rs} = \{((z_0,\ldots,z_p), (w_0,\ldots,w_q) \in S^{2p+1} \times S^{2q+1} : z_r w_s \neq 0\} .$$

V_{rs} is an open subset of $S^{2p+1} \times S^{2q+1}$. We define a map $\phi_{rs} : V_{rs} \longrightarrow \mathbb{C}^{p+q} \times T_\tau$ by

$$\phi_{rs}((z_0,\ldots,z_p),(w_0,\ldots,w_q)) = (z_0/z_r,\ldots,\widehat{z_r/z_r},\ldots,w_0/w_s,\ldots,\widehat{w_s/w_s},\ldots,t_{rs}),$$

where

$$t_{rs} = (2\pi i)^{-1}(\log(z_r) + \tau \log(w_s)), \bmod 1, \tau .$$

It is a straightforward exercise to verify that ϕ_{rs} is a homeomorphism of V_{rs} onto $\mathbb{C}^{p+q} \times T_\tau$ and that $\{(V_{rs}, \phi_{rs}) : 0 \leq r \leq p, 0 \leq s \leq q\}$ defines a complex analytic structure on $S^{2p+1} \times S^{2q+1}$. Relative to this structure $\pi : S^{2p+1} \times S^{2q+1} \longrightarrow P^p(\mathbb{C}) \times P^q(\mathbb{C})$ is holomorphic and the fibres of π are biholomorphic to T_τ. We call $S^{2p+1} \times S^{2q+1}$, with the complex structure defined above, a *Calabi-Eckmann* manifold. Observe that if $P \in S^{2p+1}$, $Q \in S^{2q+1}$ then the complex manifold $X = (S^{2p+1} \setminus \{P\}) \times (S^{2q+1} \setminus \{Q\})$ is homeomorphic to $\mathbb{C}^{p+q+1}$. However, the complex structure induced on $\mathbb{C}^{p+q+1}$ is quite different from the usual structure. We shall show that every analytic function on X is constant. To see this, notice that if

$x \in P^p(\mathbb{C}) \times P^q(\mathbb{C})$ then $(\pi|X)^{-1}(x)$ is biholomorphic to a complex torus provided that $\pi(P, Q) \neq x$. Such tori are dense in X and if $f \in A(X)$, then certainly f is constant on any of these tori (Proposition 4.2.4). Hence, by continuity, f is constant on any fibre $(\pi|X)^{-1}(x)$ and so induces a holomorphic function on $P^p(\mathbb{C}) \times P^q(\mathbb{C})$ which must be constant. Alternatively, if we use Exercise 2, §2, Chapter 2 (locally) we see that f must extend to an analytic function on $S^{2p+1} \times S^{2q+1}$ and so f must be constant. This example shows that a non-compact complex manifold which is homeomorphic to $\mathbb{C}^n$ need not be biholomorphic to any open subset of $\mathbb{C}^n$, $n \geq 3$. This is in sharp contrast to what happens in case $n = 1$.

6. *Exotic complex structures* (Brieskorn and Van de Ven [1]). Let $a = (a_0,\dots,a_n)$ be an (n+1)-tuple of strictly positive integers. Let $X(a) \subset \mathbb{C}^{n+1}$ denote the zero set of the polynomial $z_0^{a_0}+\dots+z_n^{a_n}$ and set $\sum(a) = X(a) \cap S^{2n+1}$. Now $X(a)\setminus\{0\}$ is a complex submanifold of $\mathbb{C}^{n+1}\setminus\{0\}$ and $\sum(a)$ is a differential submanifold of S^{2n+1} (for the latter statement see Milnor [2]). If we let Γ denote the cyclic subgroup of $\mathrm{Aut}(\mathbb{C}^{n+1}\setminus\{0\})$ generated by $g : (z_0,\dots,z_n) \longrightarrow (e^{1/a_0}z_0,\dots,e^{1/a_n}z_n)$, then Γ acts freely, properly discontinuously on $X(a)\setminus\{0\}$ and the resulting complex manifold $M = (X(a)\setminus\{0\})\setminus\Gamma$ is diffeomorphic to $S^1 \times \sum(a)$. If one of the a_j's equals one it is easy to see that $\sum(a)$ is diffeomorphic to S^{2n-1} and so M is a Hopf manifold. Suppose now that no a_j is equal to one. Let $\xi_j = e^{2\pi i/a_j}$, $0 \leq j \leq n$, and set

$$\Delta(t) = \prod_{0<i_k<a_k} (t-\xi_0^{i_0}\cdots\xi_n^{i_n}).$$

It is known (Brieskorn [1]) that if $n \neq 2$ $\sum(a)$ is a homotopy (2n-1)-sphere if and only if $\Delta(1) = 1$. Moreover, every homotopy (2n-1)-sphere $\sum^{2n-1}$ which bounds a parallelizable manifold is diffeomorphic to $\sum(a)$ for some (n+1)-tuple a $(n \neq 2)$. For $n \geq 4$, $\sum^{2n-1}$ is homeomorphic but not necessarily diffeomorphic to S^{2n-1} (Brieskorn [1], Milnor [3]). Now suppose $n \geq 4$ and a is chosen so that $\sum(a)$ is a homotopy sphere with non-standard differentia structure. In Brieskorn and Van de Ven [1], it is shown that $S^1 \times \sum(a)$ is not diffeomorphic to $S^1 \times S^{2n-1}$ and so the resulting complex structure on $S^1 \times \sum(a)$ is non-standard or "exotic". As a specific example, if we take $n = 4$, $a_0 = a_1 = a_2 = 2$, $a_3 = 3$ and $a_4 = 6k - 1$, it can be shown (see Brieskorn [1]) that as k varies from 1 to 28 we obtain 28 distinct complex structures on $S^1 \times S^7$, 27 of which are exotic.

Exercises.

1. In the example describing Calabi-Eckmann manifolds show that if $p = 0$, $q > 0$, the resulting manifold is biholomorphic to a Hopf manifold.

2. Let Γ be the subgroup of $GL(3, \mathbb{C})$ consisting of matrices
$$\begin{pmatrix} 1 & a & a_2 \\ 0 & 1 & a_3 \\ 0 & 0 & 1 \end{pmatrix}, \quad \text{where } a_j \in \mathbb{Z}[\sqrt{-1}] \text{ (that is, the } a_j \text{ are Gaussian integers).}$$
Prove that Γ acts freely, properly discontinuously on $\mathbb{C}^3$. The resulting complex manifold $\mathbb{C}^3/\Gamma$ is called an *Iwasawa* manifold. Note that $\pi_1(\mathbb{C}^3/\Gamma)$ is not Abelian.

§6. Analytic hypersurfaces.

In this section we make a global study of analytic sets defined locally by the vanishing of a single analytic function.

Definition 4.6.1. Let M be a complex manifold and X be a proper analytic subset of M. We say that X is an *analytic hypersurface* of M if for each $x \in X$ there exists $U \in \mathcal{U}_x$ and $f \in A(U)$ such that $X \cap U = f^{-1}(0)$.

Before stating the next Proposition we recall some notation from §5 of Chapter 3 : Suppose X is an analytic subset of M. As in §5 of Chapter 3, we may define the ideal $\mathcal{I}_x(X) \lhd \mathcal{O}_x$ at each point $x \in M$. If $x \notin X$, $\mathcal{I}_x(X) = \mathcal{O}_x$. Otherwise $\mathcal{I}_x(X)$ is a non-trivial ideal of $\mathcal{O}_x$.

Proposition 4.6.2. Let X be an analytic subset of the complex manifold M. The following conditions are equivalent.

1. X is an analytic hypersurface.
2. For each $x \in X$, $\mathcal{I}_x(X)$ is a principal ideal.

Proof. Suppose X is an analytic hypersurface and $x \in X$. Then there exists an open neighbourhood U of x and $f \in A(U)$ such that $f^{-1}(0) = X \cap U$. By the Nullstellensatz for principal ideals (Corollary 3.5.14),

$I_x(X) = \text{Rad}(f_x)$. But $f_x = \prod_{j=1}^{k} p_j^{r_j}$, where $p_j \in O_x$ are irreducible and coprime. Hence $I_x(X) = (p_1 \cdots p_k)$ and so is principal. For the converse, note that by Theorem 3.5.9 $Z(I_x(X)) = X_x$. Therefore if $I_x(X)$ is principal, say (f_x), we see that there exists an open neighbourhood U of x and representative $f \in A(U)$ of f_x such that $X \cap U = f^{-1}(0)$. □

Definition 4.6.3. Let X be an analytic hypersurface in M and $x \in X$. Suppose that $I_x(X) = (f_x)$ and that $df(x) \neq 0$ for some representative f of f_x. Then we say that x is a *regular* point of X. If x is not regular we say it is a *singular* point. We denote the set of regular and singular points of X by Reg(X) and Sing(X) respectively.

Remark. A point $x \in X$ is a regular point of X if and only if we can find an open neighbourhood V of x in M such that $V \cap X$ is a complex submanifold of M. Indeed, if the latter condition holds we may find an open neighbourhood W of x contained in V and $g \in A(W)$ such that $g^{-1}(0) = X \cap W$ and $dg(x) \neq 0$. Since $dg(x) \neq 0$ implies that g_x is irreducible, we must have $I_x(X) = (g_x)$. The converse follows from the implicit function theorem.

Theorem 4.6.4. Let X be an analytic hypersurface of the complex manifold M. We have

1. Reg(X) is an open and dense subset of X and is a codimension one complex submanifold of M.

2. Sing(X) is an analytic subset of M.

Proof. The remark above already shows that Reg(X) is an open subset of X and a codimension one complex submanifold of M.

Let $x \in X$. By Theorem 3.5.16 we may find an open neighbourhood V of x in M and $g \in A(V)$ such that $I_y(V) = (g_y)$, for all $y \in V$. Hence $\text{Sing}(X) \cap V = \{y \in V : g(y) = dg(y) = 0\}$. Hence Sing(X) is an analytic subset of M. Suppose that $\text{Reg}(X) \cap V$ is not dense in $X \cap V$. Then for some $y \in X \cap V$, there exists an open neighbourhood W of y in V such that $g(z) = dg(z) = 0$, $z \in X \cap W$. Since $I_y(X) = (g_y)$, this implies that in a local coordinate system $(z_1,\ldots,z_n)$ at y we would have

$$(\partial g/\partial z_j)_y = u_y^j g_y, \quad 1 \le j \le n,$$

where $u_y^j \in \mathcal{O}_y$. However such a relation between an analytic function and its derivatives can only hold if the function is identically zero (look at Taylor series of g and dg at y). This contradiction shows that $\mathrm{Sing}(X) \cap V$ has no interior points in $X \cap V$. Therefore, Reg(X) is dense in X. Alternatively the reader may use the results at the beginning of §5, Chapter 3 together with Corollary 3.5.15 to prove the density of Reg(X). □

The next result describes the local structure of an analytic hypersurface.

Proposition 4.6.5. Let X be an analytic hypersurface in the complex manifold M and let $x \in X$. We may find an open neighbourhood U of x in M and analytic hypersurfaces Z_j, $j = 1,\dots,k$, in U such that

1. $X \cap U = \bigcup_{j=1}^{k} Z_j$.
2. $\mathrm{Reg}(Z_j)$ is connected, $1 \le j \le k$.
3. $\mathrm{Reg}(X) \cap U = \bigcup_{j=1}^{k} (\mathrm{Reg}(Z_j) \setminus \bigcup_{l \ne j} Z_l)$.

Proof. Choose a chart (U, ϕ) for M at x such that $\phi(x) = 0$. Set $D = \phi(U)$ and $Z = \phi(X \cap U)$. Since ϕ is biholomorphic, Z is an analytic hypersurface in D. Shrinking U if necessary, we may suppose (Corollary 3.5.15 and Theorem 3.5.16) that D is a polydisc and

$$Z = \bigcup_{j=1}^{k} Z(p^j),$$

where the p^j are coprime, irreducible Weierstrass polynomials on D and $I_y(Z) = (p_y^1 \cdots p_y^k)$ for every $y \in D$. Set $Z_j = Z(p^j)$, $1 \le j \le k$. Then $\mathrm{Reg}(Z_j)$ is a connected subset of Z_j, $1 \le j \le k$ (Exercise 1, §5, Chapter 3). Since $I_y(Z) = (p_y^1 \cdots p_y^k)$, $y \in D$, we see immediately that

$$\mathrm{Reg}(Z) = \bigcup_{j=1}^{k} (\mathrm{Reg}(Z_j) \setminus \bigcup_{l \ne j} Z_l).$$

Let X_j denote the analytic hypersurface $\phi^{-1}(Z_j)$ in D, $1 \le j \le k$. Clearly $\{X_j : 1 \le j \le k\}$ satisfy the conditions of the Proposition. □

Definition 4.6.6. Let X be an analytic subset of the complex manifold M. We say that X is *reducible* if we can find analytic subsets Y, Z of M, neither of which equals X, such that $X = Y \cup Z$. If X is not reducible, we say X is *irreducible*.

Theorem 4.6.7. Let X be an analytic hypersurface of the complex manifold M and $\{X_i' : i \in I\}$ denote the set of connected components of Reg(X). Set $X_i = \bar{X}_i'$, $i \in I$. Then

1. X_i is an irreducible analytic hypersurface, $i \in I$.

2. $\{X_i : i \in I\}$ is locally finite. In particular, if M is compact I is finite.

3. $X = \bigcup_{i \in I} X_i$ and this decomposition of X as a (countable) union of irreducible analytic hypersurfaces is unique up to order.

4. If Y is any proper irreducible analytic subset of M and $Y \subset X$ then $Y \subset X_i$ for some $i \in I$.

Proof. Let $x \in X$ and suppose $x \in X_i$. We can find an open neighbourhood U of x in M and hypersurfaces Z_j, $1 \le j \le k$, in U satisfying the conditions of Proposition 4.6.5. Define

$$R_j = \text{Reg}(Z_j) \backslash \bigcup_{l \neq j} Z_l .$$

For $1 \le j \le k$, $R_j \subset \text{Reg}(X)$ and is connected. Hence for some subset $\Lambda \subset \{1,\dots,k\}$, $R_j \subset X_i'$ if and only if $j \in \Lambda$. But now $X_i \cap U = \bigcup_{j \in \Lambda} \bar{R}_i = \bigcup_{j \in \Lambda} Z_j$. This proves that X_i is an analytic hypersurface in M. Since only finitely many of the X_i can meet U, the family $\{X_i : i \in I\}$ is locall finite. Next we show that each X_i is irreducible. Suppose that $X_i = Y \cup$ where Y and Z are analytic subsets of M. Set $Y' = X_i' \cap Y$, $Z' = X_i' \cap Z$. Since X_i' is a complex submanifold we see that Y', Z' are analytic subsets of X_i'. Now Y', Z' cannot both be proper analytic subsets of X_i' since if they were they would be nowhere dense in X_i' (Corollary 2.2.3) and so X_i' could not be the union of Y' and Z'. Suppose then that $Y' = X_i'$. Taking closures we see that $Y = X_i$ and so X_i is irreducible.

Finally suppose that Y is an irreducible analytic subset and $Y \subset X$. Setting $Y_i = X_i \cap Y$, we have $Y = \bigcup_{i \in I} Y_i$. Fix $y \in Y$ and let L be the finite subset of I characterised by $j \in L$ if and only if $y \in X_j$. Since $\{X_i : I \in I\}$ is locally finite, $\sum = \bigcup_{i \in I \setminus L} Y_i$ is an analytic subset of M. Clearly $\sum \neq Y$, since $y \notin \sum$. But now we have expressed Y as a finite union

$$Y = \bigcup_{i \in L} Y_i \cup \sum$$

of analytic sets. Since $\sum \neq Y$, we must have $Y = Y_i$ for some $i \in L$. That is $Y \subset X_i$. □

As an immediate corollary of Theorem 4.6.7 we have

Theorem 4.6.8. An analytic hypersurface is irreducible if and only if its set of regular points is connected.

Remarks.

1. We call the hypersurfaces X_i constructed in Theorem 4.6.7 the *irreducible components* of X.

2. Theorems 4.6.7 and 4.6.8 hold for arbitrary analytic subsets of a complex manifold. For proofs and further details we refer the reader to Gunning and Rossi [1], R. Narasimhan [3] and Whitney [1].

3. Notice that if X is an irreducible analytic hypersurface it does not follow that X_x is irreducible at every point $x \in X$. For example, the hypersurface $y^2 = x^2(1 - x)$ is irreducible but its germ at the origin is reducible.

4. Theorems 4.6.7 and 4.6.8 do not generalise to real analytic sets. For example, $y^2 = x^3$ is irreducible in $\mathbb{R}^2$ but the set of regular points is not connected. See R. Narasimhan [3; Chapter V] for further examples of the pathology that can occur with real analytic sets.

In §2,4 of Chapter 1 we defined and discussed divisors on a Riemann surface and their relationship to meromorphic functions. For the remainder of this section we shall show how we may use Theorem 4.6.7 to give a local description of divisors on an arbitrary complex manifold.

Definition 4.6.8. Let M be a complex manifold. A *Weil divisor* on M is a formal sum $\sum_{i\in I} n_i \cdot X_i$, where $\{X_i : i \in I\}$ is a locally finite set of mutually distinct irreducible analytic hypersurfaces in M and $n_i \in \mathbb{Z}$, $i \in I$.

As in Chapter 1, we denote the set of Weil divisors on M by $\mathcal{D}(M)$ and note that $\mathcal{D}(M)$ has the structure of an ordered Abelian group. If $d = \sum_{i\in I} n_i \cdot X_i \in \mathcal{D}(M)$, we let $|d|$ denote the analytic hypersurface $\bigcup_{i\in I} X_i$.

Example 1. Let X be a non-singular analytic hypersurface in $P^n(\mathbb{C})$. Assuming Chow's theorem, we see that X is algebraic and so is the zero set of a homogeneous polynomial P. Since X is non-singular, X is (analytically) irreducible and it is not hard to verify that P can be chosen to be irreducible (as a polynomial). Let d = degree(P). We say that X is a *hypersurface of degree d*. Now suppose $d > 1$. Let $P^n(\mathbb{C})^*$ denote the set of hyperplanes in $P^n(\mathbb{C})$ (see also §4). Every hyperplane $H \in P^n(\mathbb{C})^*$ will intersect X in a Weil divisor. Indeed, changing coordinates we may assume that H is the hyperplane $z_0 = 0$. $H \cap X$ is then the algebraic set determined by $G(z_1,\ldots,z_n) = P(0,z_1,\ldots,z_n) = 0$. G may not be irreducible and we let

$$G(z_1,\ldots,z_n) = \prod_{i=1}^{q} G_i(z_1,\ldots,z_n)^{n_i}$$

denote the prime factorization of G. If we set $X_i = G_i^{-1}(0) \cap H$, we see that $X \cap H$ determines the divisor

$$d(X \cap H) = \sum_{i=1}^{q} n_i \cdot X_i .$$

Notice that $\sum_{i=1}^{q} n_i d_i = d$, where $d_i = \text{degree}(G_i)$, $1 \le i \le q$. Let $\Gamma = \{d(X \cap H) : H \in P^n(\mathbb{C})^*\} \subset \mathcal{D}(X)$. Γ is in bijective correspondence with $P^n(\mathbb{C})^*$. Given $d \in \Gamma$, we let $H(d) \in P^n(\mathbb{C})^*$ denote the corresponding hyperplane. The set of divisors Γ on X determines the embedding of X in $P^n(\mathbb{C})$. Indeed, if $x \in X$ let Γ_x denote the set of all divisors $d \in \Gamma$ such that $x \in |d|$. Define $\phi(x) \in P^n(\mathbb{C})$ to be $\bigcap_{d \in \Gamma_x} H(d)$. The reader may easily verify that ϕ embeds X in $P^n(\mathbb{C})$. As we shall see later there is a close relationship between families of divisors on compact complex manifolds and embeddings into projective space.

Our next definition is motivated by the Cousin II problem and the question of constructing a meromorphic function with specified pole and zero set with multiplicities.

Definition 4.6.9. A *Cartier divisor* d on the complex manifold M consists of an open cover $\{U_i : i \in I\}$ of M together with meromorphic functions $m_i \in M^*(U_i)$ such that for all i, $j \in I$ we have $m_i m_j^{-1} \in A^*(U_{ij})$. We write $d = \{(U_i, m_i) : i \in I\}$.

Remarks.

1. We regard the Cartier divisors $d = \{(U_i, m_i) : i \in I\}$, $d' = \{(V_j, m_j') : i \in J\}$ as being equal if $m_j' m_i^{-1} \in A^*(V_j \cap U_i)$, $i \in I$, $j \in J$. That is, we think of a Cartier divisor as an equivalence class of local data. Later, in Chapter 6, we shall give a slicker definition of Cartier divisor in terms of sheaves.

2. We let $\tilde{\mathcal{D}}(M)$ denote the set of Cartier divisors on M.

3. Of course, a Cartier divisor is nothing else than the data for the Cousin II problem (Definition 3.4.10). However, we prefer to reserve the term "Cousin II problem" for non-compact complex manifolds, especially domains of holomorphy and Stein manifolds.

Proposition 4.6.10. The set of Cartier divisors $\tilde{\mathcal{D}}(M)$ has the natural structure of an Abelian group.

Proof. Let $d = \{(U_i, m_i) : i \in I\}$, $d' = \{(V_j, m_j') : j \in J\} \in \tilde{\mathcal{D}}(M)$. We define

$$d + d' = \{(U_i \cap V_j, m_i m_j') : i \in I, j \in J\}$$

$$-d = \{(U_i, m_i^{-1}) : i \in I\}.$$

It is straightforward to verify that these definitions do not depend on the particular choice of local data for the divisors. □

Example 2. We have a natural group homomorphism $\mathrm{div} : M^*(M) \longrightarrow \tilde{\mathcal{D}}(M)$ defined by $\mathrm{div}(m) = \{(M, m)\} \in \tilde{\mathcal{D}}(M)$. $\mathrm{div}(m)$ is called the divisor of m.

Theorem 4.6.11. There is a natural group isomorphism between $\mathcal{D}(M)$ and $\tilde{\mathcal{D}}(M)$.

Proof. Let $d = \sum_{i\in I} n_i \cdot X_i \in \mathcal{D}(M)$. Given $x \in M$, choose an open neighbourhood U_x of x and $p^i \in A(U_x)$, $i \in I$, such that

a) All but finitely many of the p^i are identically equal to 1.

b) $I_y(X_i) = (p^i_y)$, $y \in U_x$, $i \in I$.

$$\text{Set } m_x = \prod_{i\in I} (p^i)^{n_i} \in M^*(U_x).$$

Clearly $\phi(d) = \{(U_x, m_x) : x \in M\}$ is a Cartier divisor on M. The map $\phi : \mathcal{D}(M) \longrightarrow \tilde{\mathcal{D}}(M)$ is easily seen to be an injective group homomorphism.

Conversely, suppose $d = \{(U_i, m_i) : i \in I\}$ is a Cartier divisor on M. Define $X = \bigcup_{i\in I} (Z(m_i) \cup P(m_i))$. X is a well defined analytic hypersurface in M. Indeed, $X \cap U_i = Z(m_i) \cup P(m_i)$ since the condition $m_i m_j^{-1} \in A^*(U_{ij})$ implies that m_i and m_j have the same pole and zero sets on U_{ij}. Let $\{X_\alpha : \alpha \in \Lambda\}$ be the decomposition of X into its irreducible components. Refining the cover $\{U_i : i \in I\}$ if necessary, we may find for each $i \in I$, $p_\alpha \in A(U_i)$ such that

a) All but finitely many of the p_α are identically equal to 1.

b) $U_i \cap X_\alpha = p_\alpha^{-1}(0)$, $\alpha \in \Lambda$.

c) $I_y(X_\alpha) = (p_{\alpha,y})$, $y \in U_i$, $\alpha \in \Lambda$.

Given $y \in U_i$, we see that there exist a unit $u_y \in \mathcal{O}_y$ and integers n_α, $\alpha \in \Lambda$, such that

$$m_y = u_y \prod_{\alpha\in\Lambda} p_{\alpha,y}^{n_\alpha}.$$

Hence for some $u \in A^*(U_i)$, we have $m_i = u \prod_{\alpha\in\Lambda} p_\alpha^{n_\alpha}$. (We assume here that U_i is connected, otherwise we could only claim that the n_α were constant on connected components of U_i).

We claim that the integers n_α do not depend on $i \in I$. This follows since n_α is constant on the connected open set $\text{Reg}(X_\alpha) \cap \text{Reg}(X)$. Since $\text{Reg}(X_\alpha) \cap \text{Reg}(X)$ is dense in X_α, n_α is constant on X_α. We now define

$$\gamma(d) = \sum_{\alpha\in\Lambda} n_\alpha \cdot X_\alpha \in \mathcal{D}(M).$$

By our construction it is clear that $\gamma = \phi^{-1}$ and so $\mathcal{D}(M)$ and $\tilde{\mathcal{D}}(M)$ are naturally isomorphic. □

Remarks.

1. In the sequel we regard $\mathcal{D}(M)$ and $\tilde{\mathcal{D}}(M)$ as identified by the isomorphism constructed above and write $\mathcal{D}(M)$ for the set of divisors on M, omiting the prefix "Weil" or "Cartier".

2. If we attempt to define divisors on more general objects, such as analytic sets with singularities, we find that the sets of Weil and Cartier divisors need not agree and in practice we tend to work with the less geometrical Cartier divisors (see Hartshorne [2; pages 140-142]).

Example 3. Let $m \in M^*(P^n(\mathbb{C}))$. Then $\text{div}(m) = \sum_{i=1}^{q} n_i \cdot X_i$, where the X_i are irreducible analytic hypersurfaces in $P^n(\mathbb{C})$. By Chow's theorem and Example 1, each X_i is an irreducible algebraic hypersurface of degree d_i, say. Suppose that X_i is the zero locus of the irreducible homogeneous polynomial P_i of degree d_i, $1 \le i \le q$. Define $R = \prod_{i=1}^{q} P_i^{n_i}$. Regarding m and R as defining meromorphic functions on $\mathbb{C}^{n+1}\setminus\{0\}$, we clearly have $\text{div}(m) = \text{div}(R)$ and so $\text{div}(R^{-1}m) = 0$. But therefore $R^{-1}m$ is an analytic function on $\mathbb{C}^{n+1}\setminus\{0\}$ and so extends by Hartog's theorem to an analytic function on the whole of $\mathbb{C}^{n+1}$. Taking the Taylor expansion of $R^{-1}m$ at 0 and noting that m is homogeneous of degree zero, we see that $R^{-1}m$ is a homogeneous polynomial of degree $-\sum_{i=1}^{q} n_i d_i$. Hence $\sum_{i=1}^{q} n_i d_i = 0$

(otherwise $\operatorname{div}(R^{-1}m)$ could not vanish). Consequently, $R^{-1}m = c$, for some $c \in \mathbb{C}^{\bullet}$ and so

$$m = c \prod_{i=1}^{q} P_i^{n_i} \quad \text{on} \quad P^n(\mathbb{C}) .$$

We have shown that every meromorphic function on $P^n(\mathbb{C})$ is rational. Our arguments further prove that a divisor $d = \sum_{i=1}^{q} n_i \cdot X_i$ on $P^n(\mathbb{C})$ is the divisor of a meromorphic (rational) function on $P^n(\mathbb{C})$ if and only if $\sum_{i=1}^{q} n_i d_i = 0$, where d_i is the degree of the hypersurface X_i. Given $d = \sum_{i=1}^{q} n_i \cdot X_i \in \mathcal{D}(P^n(\mathbb{C}))$, we may define the degree of d, deg(d), to be the sum $\sum_{i=1}^{q} n_i d_i$, where d_i is the degree of the hypersurface X_i. Degree defines a homomorphism $\deg : \mathcal{D}(P^n(\mathbb{C})) \longrightarrow \mathbb{Z}$ and the kernel of the degree map is precisely the set of divisors of rational functions on $P^n(\mathbb{C})$. (For an alternative proof of the rationality of meromorphic functions on $P^n(\mathbb{C})$, avoiding the use of Chow's theorem, see Jackson[1]).

§7. Blowing up

In this section we describe an operation on complex manifolds that is of the greatest importance in the study of singularities of analytic sets and the classification theory of complex manifolds.

Let $\Sigma \subset \mathbb{C}^n \times P^{n-1}(\mathbb{C})$ be the set of points $((z_1,\ldots,z_n),(\zeta_1,\ldots,\zeta_n))$ satisfying the equations

$$z_i\zeta_j = z_j\zeta_i, \quad 1 \le i, j \le n .$$

(We suppose in what follows that $n \ge 2$). Notice that $((z_1,\ldots,z_n), (tz_1,\ldots,tz_n)) \in \Sigma$, for all $(z_1,\ldots,z_n) \neq 0$ and $t \in \mathbb{C}^{\bullet}$. In other words, given $X \in \mathbb{C}^n$, $X \neq 0$, Σ contains the point corresponding to X and the line through X. Since $\{0\} \times P^{n-1}(\mathbb{C}) \subset \Sigma$, we see that Σ contains the point corresponding to zero and all the lines through zero. The reader may easily verify that we have described all the points in Σ. Let $\pi : \Sigma \longrightarrow \mathbb{C}^n$ denote the restriction of the projection of $\mathbb{C}^n \times P^{n-1}(\mathbb{C})$ on $\mathbb{C}^n$ to Σ. If we set $E = \pi^{-1}(0)$, then E is biholomorphic to $P^{n-1}(\mathbb{C})$ and π

maps $\Sigma \setminus E$ bijectively onto $\mathbb{C}^n \setminus \{0\}$. We call Σ the *blowing up* of $\mathbb{C}^n$ at 0. Amongst many other commonly used terms to describe Σ are : $\mathbb{C}^n$ blown up at zero; the *quadratic transform* of $\mathbb{C}^n$ at zero; the *monoidal transform* of $\mathbb{C}^n$ at zero; the *Hopf* σ*-process* of $\mathbb{C}^n$ at zero. We call 0 the *centre* of the blowing up; E is called the *exceptional variety*.

It is easy to see that Σ is a complex submanifold of $\mathbb{C}^n \times P^{n-1}(\mathbb{C})$ and we shall now construct an explicit atlas for Σ . For $1 \leq j \leq n$, define $\gamma_j : \mathbb{C}^n \longrightarrow \Sigma \subset \mathbb{C}^n \times P^{n-1}(\mathbb{C})$ by

$$\gamma_j(X_1,\ldots,X_n) = ((X_1X_j,\ldots,X_{j-1}X_j,X_j,X_{j+1}X_j,\ldots,X_nX_j),(X_1,\ldots,X_{j-1},1,\ldots,X_n)).$$

The reader may verify that γ_j maps $\mathbb{C}^n$ homeomorphically onto an open subset of Σ . If we set $U_j = \gamma_j(\mathbb{C}^n)$, $\phi_j = \gamma_j^{-1}$, then $\{(U_j, \phi_j), 1 \leq j \leq n\}$ is a complex analytic atlas on Σ . Relative to this complex structure on Σ, $\pi : \Sigma \longrightarrow \mathbb{C}^n$ is holomorphic and π maps $\Sigma \setminus E$ biholomorphically onto $\mathbb{C}^n \setminus \{0\}$.

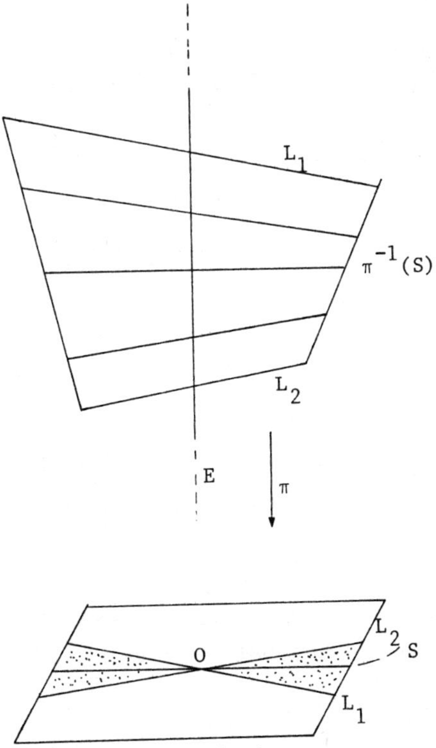

Figure 8 *Blowing up*

In the figure the cone $S \subset \mathbb{C}^n$ is blown up into an open subset of Σ. That is, the cone is "untwisted" into an open set. Notice how lines passing through the origin of $\mathbb{C}^n$ lift to distinct lines in Σ. We can also describe this phenomenon using charts of the atlas we constructed above for Σ. Thus if, $n = 2$ and we take any open cone $S \subset \mathbb{C}^2$ which contains the z_1-axis but omits the z_2-axis then $\pi^{-1}(S)$ is described by the figure below if we use coordinates given by the chart (U_1, ϕ_1).

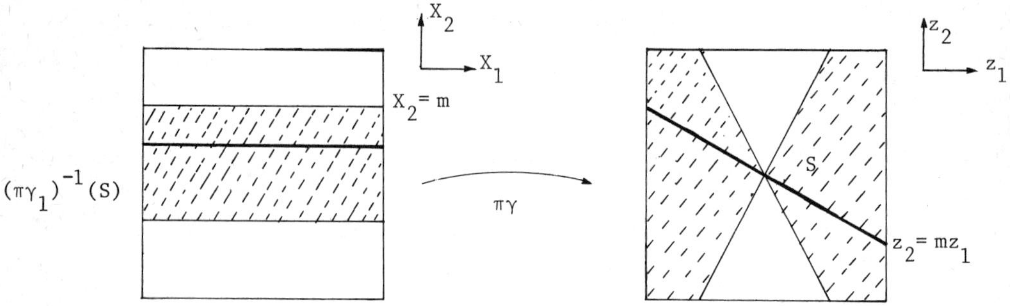

Figure 9

In the figure, the line $z_2 = mz_1$ in the (z_1, z_2)-plane corresponds to the affine line $X_2 = m$ in the (X_1, X_2)-plane.

The construction we have described above works equally well if instead of $\mathbb{C}^n$ we take an open neighbourhood U of $0 \in \mathbb{C}^n$ and define $\Sigma_U \subset U \times P^{n-1}(\mathbb{C})$ to be the intersection of Σ with $U \times P^{n-1}(\mathbb{C})$. We refer to Σ_U as U blown up at zero. The projection again restricts to a holomorphic map $\pi : \Sigma_U \longrightarrow U$ which induces a biholomorphic map between $\Sigma_U \setminus E$ and $U \setminus \{0\}$.

We now wish to generalise the construction above so that we can blow up arbitrary complex manifolds at a point. First, a lemma.

Lemma 4.7.1. Let M and Σ be complex manifolds with closed submanifolds N and X respectively. Suppose that for some open neighbourhood U of N in M there exists a biholomorphism $\phi : U \setminus N \longrightarrow \Sigma \setminus X$. Then if we define

$$M^* = (M \setminus N) \cup \Sigma,$$

where we identify $x \in M \setminus N$ with $y \in \Sigma \setminus X$ if $\phi(x) = y$, M^* has the natural structure of a complex manifold such that $M^* \setminus X$ is biholomorphic to $M \setminus N$. Furthermore, if ϕ is the restriction of a holomorphic map $\pi : U \longrightarrow \Sigma$, then π induces a holomorphic map $\pi : M^* \longrightarrow M$ such that π maps $M^* \setminus X$ biholomorphically onto $M \setminus N$.

Proof. The proof is quite elementary and we leave it to the reader. □

Suppose M is a complex manifold of dimension n, $n > 1$, and let $p \in M$. Choose a coordinate chart (V, ϕ) for M such that $p \in V$ and $\phi(p) = 0$ Set $U = \phi(V)$. Let Σ denote the blow up of U at 0 (Σ_U in the notation above). Then $\phi^{-1}\pi : \Sigma \setminus E \longrightarrow V \setminus \{p\}$ is a biholomorphism and so we may apply Lemma 4.7.1 to construct the complex manifold

$$B_p(M) = (M \setminus \{p\}) \cup \Sigma$$

together with the holomorphic map $\pi : B_p(M) \longrightarrow M$. We say that $B_p(M)$ is M blown up at the point p (or any of the other descriptions we gave previously for blowing up $\mathbb{C}^n$). We call p the centre of the blowing up and $E = \pi^{-1}(p)$ the exceptional variety. Clearly E is biholomorphic to $P^{n-1}(\mathbb{C})$. The projection map $\pi : B_p(M) \longrightarrow M$ is a proper holomorphic map and restricts to a biholomorphism of $B_p(M) \setminus E$ with $M \setminus \{p\}$.

We may generalise the above construction to include centres which are closed complex submanifolds of M. First we describe the local situation. Regard $\mathbb{C}^m$ as the subspace $\mathbb{C}^m \times \{0\}$ of $\mathbb{C}^{m+n}$. We define $\mathbb{C}^{m+n}$ blown up along $\mathbb{C}^m$ to be the space $\mathbb{C}^m \times \Sigma$, where Σ is $\mathbb{C}^n$ blown up at zero. In other words, we just blow up normal to $\mathbb{C}^m$ in $\mathbb{C}^{m+n}$. More generally, if X is a closed submanifold of M we may blow up M along X to construct a complex manifold $B_X(M)$ together with a holomorphic projection $\pi : B_X(M) \longrightarrow M$ which restricts to a biholomorphic map of $B_X(M) \setminus \pi^{-1}(X)$ onto $M \setminus X$. In this case the exceptional variety $\pi^{-1}(X)$ is a bundle over X with fibre biholomorphic to $P^q(\mathbb{C})$, where $q = \dim(M) - \dim(X) - 1$. We give an example of this process below.

We shall now give some examples to show how blowing up may be used to "desingularize" analytic sets. Suppose that X is an analytic subset of the complex manifold M. Let Y be a closed complex submanifold of M which is a subset of Sing(X). Let us blow up M along Y to give the new complex manifold $B_Y(M)$. The set $\pi^{-1}(X)$ is an analytic subset of $B_Y(M)$. Necessarily $\pi^{-1}(X)$ contains the exceptional variety $E = \pi^{-1}(Y)$ and we define the *strict transform* of X to be the analytic subset $X^* = \overline{\pi^{-1}(X)\setminus E}$ of $B_Y(M)$. As our examples will show a careful choice of Y will often result in X^* having "simpler" singularities than X.

Examples.

1. Let $Z \subset \mathbb{C}^2$ denote the curve defined by $P(z_1, z_2) = (z_1-z_2)(z_1+z_2)=0$. Z has an isolated singularity at zero. As above we let $\sum$ denote $\mathbb{C}^2$ blown up at zero and $\pi : \sum \longrightarrow \mathbb{C}^2$ the projection map. Set $\tilde{Z} = \pi^{-1}(Z)$ and let Z^* denote the strict transform of Z. To describe $\tilde{Z}$ we use the atlas $\{(U_i, \phi_i) : i = 1, 2\}$ that we previously constructed on $\sum$. For $i = 1, 2$, set $\tilde{Z}_i = \phi_i(\tilde{Z} \cap U_i) \subset \mathbb{C}^2$. Then $\tilde{Z}_1$ is the zero set of $P(X_1, X_1X_2)$ and $\tilde{Z}_2$ is the zero set of $P(X_1X_2, X_2)$. We have

$$P(X_1, X_1X_2) = (X_1 - X_1X_2)(X_1 + X_1X_2) = X_1^2(1 - X_2)(1 + X_2) .$$

Now $X_1 = 0$ is just the equation of the exceptional variety (intersected with U_1) and so the strict transform Z^* of Z is the pair of distinct lines $X_2 = \pm 1$ (again intersected with U_1). A similar description holds in the chart (U_2, ϕ_2). Hence Z^* consists of two distinct lines and, in particular, is non-singular.

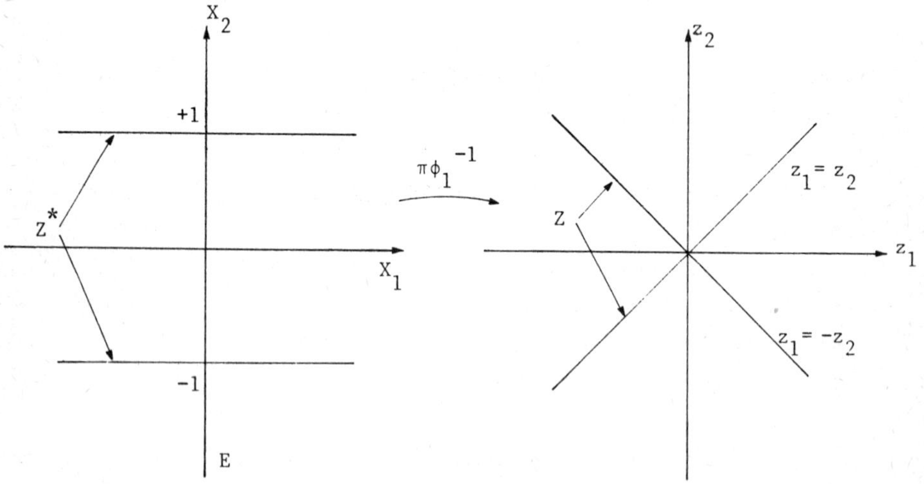

Figure 10

2. Let $Z \subset \mathbb{C}^2$ be the curve defined by $P(z_1, z_2) = z_1^2 - z_2^3 = 0$. Z has an isolated singularity at zero. Repeating the argument above, we see that $\tilde{Z}_1 \subset \mathbb{C}^2$ is the zero locus of $X_1^2 - X_1^3X_2^3 = X_1^2(1 - X_1X_2^3)$. That is, the intersection of Z^* with U_1 is the non-singular curve $1 = X_1X_2^3$ (notice that this curve does not meet the exceptional variety $X_1 = 0$ – in U_1). Similarly, $\tilde{Z}_2$ is the zero locus of $X_1^2X_2^2 - X_2^3 = X_2^2(X_1^2 - X_2)$. Now $X_2 = 0$ is the equation of the exceptional variety in U_2 and so the intersection of Z^* with U_2 is the parabola $X_2 = X_1^2$, which is of course non-singular. Hence Z^* is non-singular.

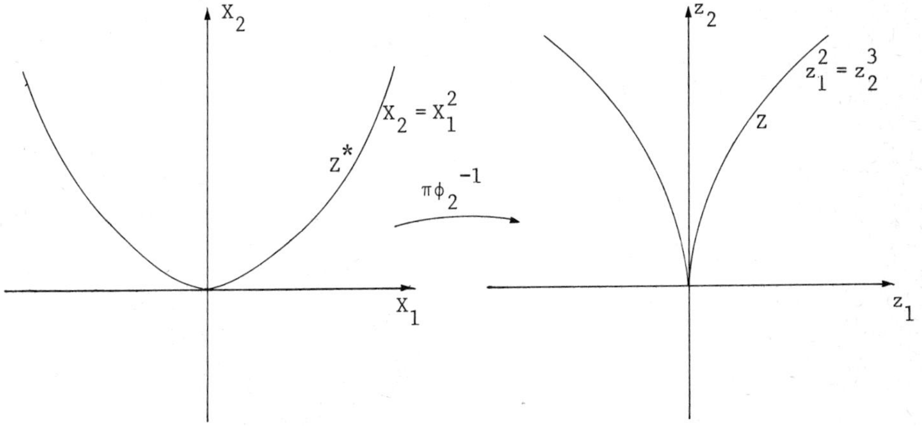

Figure 11

If instead we had started with the curve $z_1^2 = z_2^5$, we would find that $Z^* \cap U_2$ was the curve $X_1^2 = X_2^3$. Hence a further blowing up, this time of the manifold $\sum$, would remove the singularity. Without too much difficulty this technique will effectively resolve the singularities of all curves. Details may be found in Walker [1], Mumford [1] or Hartshorne [2].

Fundamental results of Hironaka [1, 2], following on work of Walker and Zariski assert that we can always resolve singularities of complex analytic sets. Specifically, if X is a complex analytic subset of a complex manifold M, there exists a (locally finite) sequence

$$M_\Lambda \xrightarrow{\pi_\Lambda} M_{\Lambda-1} \xrightarrow{\pi_{\Lambda-1}} \cdots \longrightarrow M_1 \xrightarrow{\pi_1} M$$

of blowing ups of M, with non-singular centres, such that the iterated strict transform of X is a non-singular submanifold of $M_\wedge$. The proof of this result is extremely hard. Much of the difficulty lies with choosing the centres so that when we blow up along them the singularities are somehow simplified. A proof of the desingularization theorem, together with many helpful examples is given in Hironaka [2].

The next example shows that one cannot hope to give simple proofs of the desingularization theorem by working with a restricted class of analytic sets such as hypersurfaces with isolated singularities.

Example 3. Let $Z \subset \mathbb{C}^3$ be the hypersurface defined by $z_1^5 = z_2 z_3^6 - z_2^6 z_3$. Z has an isolated singularity at 0. Blow up $\mathbb{C}^3$ at zero. We now describe $\tilde{Z}_i = \tilde{Z} \cap U_i$, $1 \leq i \leq 3$.

$\tilde{Z}_1$ is the zero locus of $X_1^5(1 - X_1^2 X_2 X_3^6 + X_1^2 X_2^6 X_3)$ and so the strict transform Z^* intersects U_1 in the non-singular hypersurface $1 = X_1^2 X_2 X_3^6 - X_1^2 X_2^6 X_3$.

$\tilde{Z}_2$ is the zero locus of $X_2^5(X_1^5 - X_2^2 X_3^6 - X_2^2 X_3)$. Therefore $Z^* \cap U_2$ is the hypersurface $X_1^5 = X_2^2 X_3^6 - X_2^2 X_3$. Now $\mathrm{Sing}(Z^* \cap U_2)$ is the set $X_1 = X_2 = 0$. In particular, $Z^* \cap U_2$ no longer has an isolated singularity.

Just as for $\tilde{Z}_2$, we find that $Z^* \cap U_3$ is the curve $X_1^5 = X_3^2 X_2 - X_3^2 X_2^6$ with singular locus $X_1 = X_3 = 0$.

It follows from the above that $\mathrm{Sing}(Z^*)$ is a subset of the exceptional variety $P^2(\mathbb{C})$ and is in fact the hyperplane $X_1 = 0$ (Here we take homogeneous coordinates (X_1, X_2, X_3) on $P^2(\mathbb{C})$). In other words $\mathrm{Sing}(Z^*)$ is biholomorphic to the Riemann sphere $P^1(\mathbb{C})$.

We now blow up $\sum$ along $\mathrm{Sing}(Z^*)$. Let us work in U_2. $Z^* \cap U_2$ is the hypersurface $X_1^5 = X_2^2 X_3^6 - X_2^2 X_3$ with singular set $X_1 = X_2 = 0$. Blowing up along the X_3-axis, we make the transformations $X_1 = Y_1 Y_2$, $X_2 = Y_2$, $X_3 = Y_3$ to obtain the hypersurface

$$Y_1^5 Y_2^5 - Y_2^2 Y_3^6 + Y_2^2 Y_3 = Y_2^2(Y_1^5 Y_2^3 - Y_3^6 + Y_3) = 0 .$$

The strict transform is the hypersurface $Y_3 = Y_3^6 - Y_1^5 Y_2^3$ which is of course non-singular. A similar result holds in U_3 and so the strict transform of Z^* obtained by blowing up $\sum$ along $\mathrm{Sing}(Z^*)$ is a non-singular hypersurface.

We now wish to say a few words about the *embedded resolution of singularities*. First we give a definition : Suppose that Y is an analytic hypersurface in the complex manifold N. For each $x \in Y$, we may write $I_x(Y) = (p_1,\ldots,p_k)$, where $p_1,\ldots,p_k$ are the local defining equations for the irreducible components of Y passing through x. We say that Y is a *hypersurface with* (or *in*) *normal crossings* if (1) Each irreducible component of Y is non-singular; (2) For each $x \in Y$, $\{dp_1(x),\ldots,dp_k(x)\}$ form a linearly independent set.

Notice that the second condition implies that not more than dimension(N) irreducible components of Y pass through any point of N. Condition (2) may also be given in terms of the tangent spaces to the irreducible components of Y at x. We require $\dim(T_x N \cap \bigcap T_x Y_j) = n - k$, where we assume that there are k irreducible components Y_j of Y passing through x. Yet another equivalent formulation is that given $x \in Y$, we can choose complex analytic coordinates $(z_1,\ldots,z_n)$ at x such that Y is locally the zero set of the function $u(z)z_1^{m_1}\cdots z_k^{m_k}$, $u(0) \neq 0$.

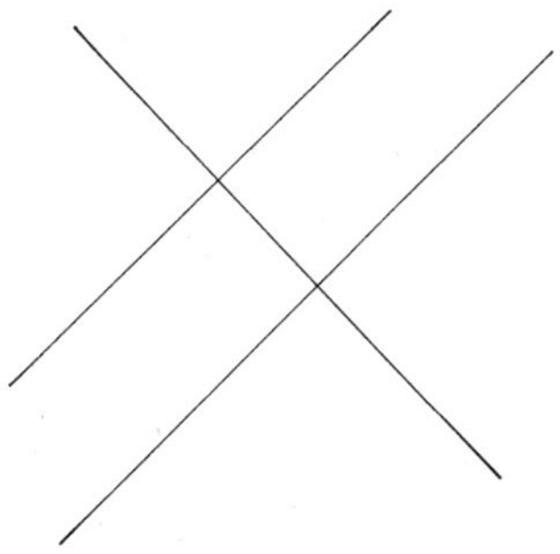

Hypersurface in normal crossings

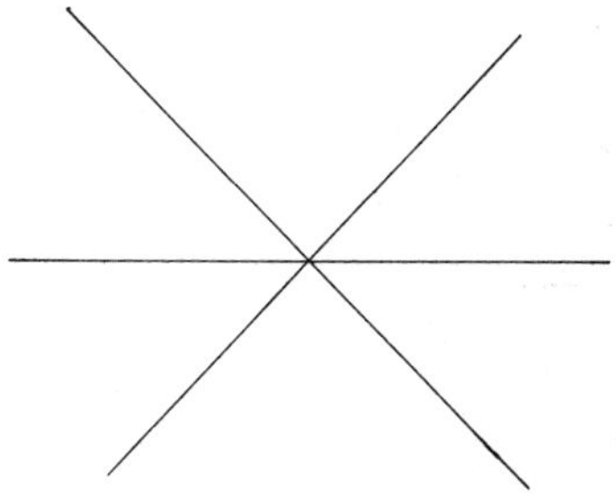

Not in normal crossings

Figure 12

Suppose $d \in \mathcal{D}(N)$. We say that d is a *divisor in normal crossings* if the hypersurface $|d|$ is in normal crossings. If d is in normal crossings and $x \in |d|$, we may choose complex analytic coordinates at x so that d is locally given as the divisor of the meromorphic function $u(z)z_1^{m_1} \cdots z_k^{m_k}$, $u(0) \neq 0$.

Now suppose X is an analytic subset of the complex manifold M. The embedded resolution of singularities theorem asserts that we can find a proper map $\pi : \tilde{M} \longrightarrow M$ of complex manifolds which is a (locally finite) composition of blowing ups with non-singular centres such that $\tilde{X} = \pi^{-1}(X)$ is non-singular and $\pi^{-1}(\mathrm{Sing}(X))$ is a divisor in $\tilde{X}$ in normal crossings. A proof may be found in Hironaka [2].

Examples.

4. The resolution of the curve $(z_1 - z_2)(z_1 + z_2) = 0$ given in example 1 above is an embedded resolution.

5. The resolution of the curve $z_1^2 - z_2^3 = 0$ given in example 2 above is not an embedded resolution. Two further blowing ups are needed to obtain an embedded resolution :

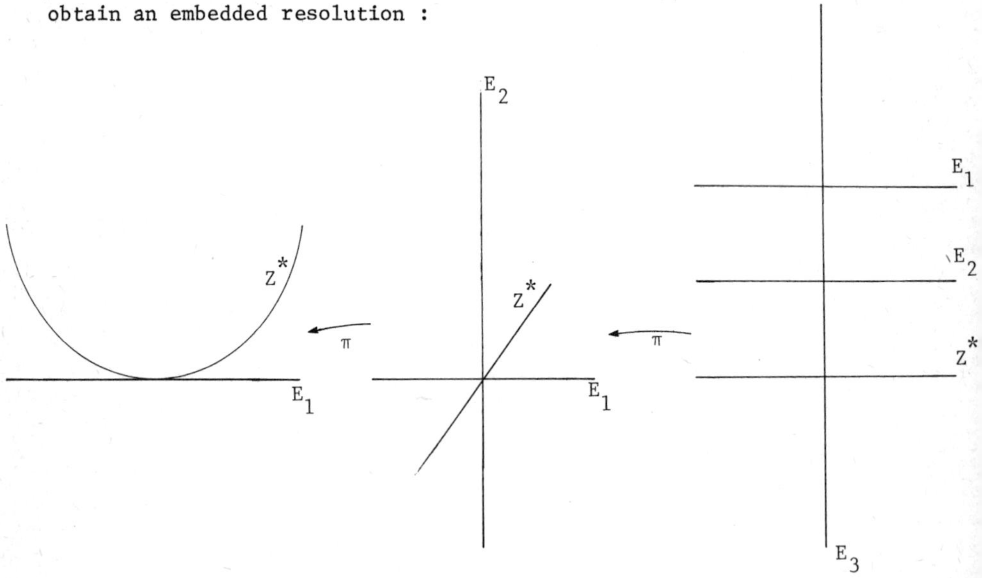

E_j denotes the exceptional curve of the jth blowing up

Figure 13

We conclude this chapter with some remarks on the role of blowing up in the classification theory of compact complex manifolds, especially complex surfaces.

First notice that we can use blowing up to define a relation between complex manifolds of the same dimension. Thus we shall say that M is B-related to N if M may be obtained by blowing up N a finite number of times. If M and N are complex surfaces then the centres of the blowing ups must always be points and the exceptional varieties are always biholomorphic to $P^1(\mathbb{C})$. Now suppose that E is a complex submanifold of the complex surface M which is biholomorphic to $P^1(\mathbb{C})$. It follows from a theorem of Grauert [1], that if the self-intersection number of E is -1 ("$E^2 = -1$") then we can *blow down* E to a point. That is, there exists a complex surface N such that for some $p \in N$, $B_p(N) = M$ and the exceptional variety of the blowing up is E. We call such a variety E an *exceptional curve of first kind.* We can now define an equivalence relation between compact complex surfaces. We say that the surfaces M and N are B-equivalent if M can be obtained from N by a finite number of blowing ups and blowing downs. We say that a complex surface is a *relatively minimal model* if and only if it does not contain any exceptional curves of the first kind. Every compact complex surface can be obtained from a relatively minimal model by a finite succession of blowing ups.

In Kodaira [2] there is a description of the relatively minimal models of compact complex surfaces as well as a description of the classification of compact complex surfaces. A useful survey of Kodaira's work, together with further references, may be found in Sundararaman [1]. See also Ueno [1].

It is an important problem to find properties of complex manifolds invariant under blowing up or blowing down (in algebraic geometry the corresponding problem is that of finding birational invariants). We shall describe one invariant here (see also §9, Chapter 5) and refer the reader to the references cited above for the description of other invariants and proofs. Let M be a compact complex manifold with field of meromorphic functions $\mathcal{M}(M)$. We let t(M) denote the transcendence degree of $\mathcal{M}(M)$ over $\mathbb{C}$. A result of Thimm [1] (see also Remmert [1], Siegel [2]) asserts that

$t(M) \leq$ dimension(M). If M is algebraic, $t(M) =$ dimension(M). We have the important result that $t(M)$ is invariant under blowing ups. Thus surfaces with different trancendence degrees cannot have the same relatively minimal models. If $t(M) =$ dimension(M) we say that M is a *Moishezon* manifold. If dimension(M) $= 2$, every Moishezon manifold is algebraic by a theorem of Kodaira [2]. The theory of Moishezon manifolds is expounded in Moishezon [1]. We remark that even though a Moishezon manifold M need not be algebraic (at least if dimension(M) > 2), there exists a finite sequence of blowing ups of M such that the resulting manifold is algebraic (For a proof see Moishezon [1, Chapter 2]). Of course, if M is algebraic, then any finite sequence of blowing ups of M with non-singular centres (necessarily algebraic) will give an algebraic manifold.

Exercises.

1. Blow up the surface $zx^2 = y^2$ at zero. Observe that blowing up at the (geometrically) most singular points does not always work. Now find a desingularization of the surface.

2. Verify that the surface $x^2y^2 - z^6 + x^6 + y^6 = 0$ has an isolated singularity at zero. Show that when we blow up at zero, the new surface has a singular set with singularities.

Bibliography

Abraham, R. and Marsden, J.E.

1. Foundations of Mechanics, 2nd. Edition, Bejamin/Cummings, Reading (1978).

Ahlfors, L.V.

1. Complex analysis, McGraw-Hill, New York (1966).
2. The complex analytic structure of the space of closed Riemann surfaces, Analytic Functions, Princeton University Press, Princeton (1960).

Ahlfors, L.V. and Sario, L.

1. Riemann Surfaces, Princeton University Press, Princeton (1960).

Aizenberg, L.A.

1. Integral representations of functions which are holomorphic in convex regions of C^n space, Sov. Math. Dokl., 4 (1963), 1149-1152.

Alexander, H.

1. Holomorphic mappings from the ball and polydisc, Math. Ann. 209 (1974), 249-256.

Andreotti, A. and Frankel, T.

1. The Lefschetz theorem on hyperplane sections, Ann. Math. 69 (1959), 713-717.

Andreotti, A. and Grauert, H.

1. Théorèmes de finitude pour la cohomologie des espaces complexes, Bull. Soc. Math. France, 90 (1962), 193-259.

Atiyah, M.

1. K theory, W.A. Benjamin Inc., Reading, Mass. (1967).

Behnke, H. and Stein, K.

1. Entwicklung analytischer Functionen auf Riemannschen Flächen, Math. Ann. 120 (1948), 430-461.

Bergman, S.

1. Sur les fonctions orthogonales de plusieurs variables complexes avec les applications à la théorie des fonctions analytiques, Mémor. Sci. Math. 106, Gauthier-Villars, Paris (1947).
2. The Kernel Function and Conformal Mapping, A.M.S. Math. Surveys, V, New York (1950).

Bers, L.

1. Uniformization, Moduli and Kleinian groups, Colloquium Lectures, Amer. Math. Soc. (1971).

Bierstone, E. and Schwartz, G.W.

1. Continuous linear division and extension of C^∞ functions, to appear.

Bishop, E.

1. Mappings of partially analytic spaces, Amer. J. Math., 83 (1961), 209-242.

Bochner, S.

1. Analytic and meromorphic continuation by means of Green's formula, Ann. Math., 49 (1943), 652-673.

Bonnesen, T. and Fenchel, W.

1. Theorie der Konvexen Körper, Ergenbisse der Math., Berlin (1934).

Bremermann, H.J.

1. Über die Äquivalenz der pseudoconvexen Gebiete und der Holomorphiegebiete in Raume von n komplexen Veranderlichen, Math. Ann., 128 (1954), 63-91.
2. Complex convexivity, Trans. Amer. Math. Soc., 82 (1956), 17-51.
3. Die holomorphiehullen der Tuben und Halblubengebiete, Math. Ann., 127 (1954).

Brieskorn, E.

1. Beispiele zur Differential topologie von Singularitäten, Invent. Math., 2 (1966), 1-14.

Brieskorn, E. and Van der Ven, A.

1. Some complex structures on products of homotopy spheres, Topology, 7 (1968), 389-393.

Calabi, E. and Eckmann, B.

1. A class of compact complex manifolds which are not algebraic, Ann. Math., 58 (1953), 494-500.

Cartan, E.

1. Sur les domaines bornés, homogènes de l'espace de n variables complexes, Abh. Math. Sem. Univ. Hamburg, 11 (1936), 116-162.

Cartan, H.

1. Les problèmes de Poincaré et de Cousin pour les fonctions de plusieurs variables complexes, C.R. Acad. Sci. Paris, 199 (1934), 1284-1287.
2. Seminaire Henri Cartan, 1951/52, W.A. Benjamin Inc., New York (1967).
3. Seminaire Henri Cartan, 1953/54, W.A. Benjamin Inc., New York (1967).
4. Sur les groupes de transformations analytiques, Act. Sc. et Ind., Hermann, Paris (1935).

Chern, S.S.

1. Complex manifolds without potential theory, Van Nostrand, Princeton (1967).

Chern, S.S. and Moser, J.

1. Real hypersurfaces in complex manifolds, Acta. Math., 133 (1974), 219-271.

Chillingworth, D.R.J.

1. Differential Topology with a view to applications, Pitman, London (1976).

Cornalba, M.

1. Complex Tori and Jacobians, Complex Analysis and its Applications, 1, IAEA, Vienna (1976), 39-100.

De La Harpe, P.

1. Introdution to Complex Tori, Complex Analysis and its Applications, 1, IAEA, Vienna (1976), 101-144.

Diederich, K.

1. Some recent developments in the theory of the Bergman kernel function, A survey, Proc. A.M.S. Symp. in Pure Math., XXX, 1, Providence (1977), 127-138.

Dieudonné, J.

1. Foundations of Modern Analysis, Academic Press, New York, (1960).

Douady, A.
1. Le problème des modules pour les sous-espaces analytiques compact d'un espace analytique donné, Ann. Inst. Fourier, Univ. de Grenoble, XVI (1966), 11-95.
2. Flatness and Privilege, Monographie 17 de l'Enseignment Mathematique, Geneva (1968), 47-74.

Ehrenpreis, L.
1. A new proof and an extension of Hartog's theorem, Bull. Amer. Math. Soc., 67 (1961), 507-509.

Field, M.J.
1. Differential Calculus and its Applications, Van Nostrand, London (1976).

Frankel, T.
1. Manifolds with positive curvature, Pac. J. Math., 11 (1961), 165-174.

Fuks, B.A.
1. Special Chapters in the Theory of Analytic Functions of Several Complex Variables, A.M.S. translations of mathematical monographs, 14 (1965).
2. Analytic Functions of Several Complex Variables, A.M.S. translations of mathematical monographs, 8 (1963).

Gamelin, T.W.
1. Uniform Algebras, Prentice-Hall, Englewood-Cliffs (1970).

Grauert, H.
1. Über Modifikation und exzeptionelle analytische Mengen, Math. Ann., 146 (1962), 331-368.

Grauert, H. and Fritzche, K.
1. Several Complex Variables, Graduate Texts in Mathematics, Springer-Verlag, New York (1976).

Grauert, H. and Remmert, R.
1. Theory of Stein spaces, Springer-Verlag, New York (1979).

Griffiths, P.
1. Periods of integrals on algebraic manifolds: summary of main results and discussion of open problems, Bull. Amer. Math. Soc., 76 (1970), 228-296.

Griffiths, P. and Harris, J.
1. Principles of Algebraic Geometry, Wiley-Interscience, J. Wiley and Sons, New York (1978).

Gunning, R.C.
1. Lectures on Riemann Surfaces, Princeton Math. notes, 12, Princeton University Press (1966).

Gunning, R.C. and Rossi, H.
1. Analytic Functions of Several Complex Variables, Prentice-Hall, Englewood-Cliffs, New Jersey (1965).

Hartshorne, R.
1. Ample Subvarieties of Algebraic Varieties, Lecture notes in Mathematics 156, Springer-Verlag (1970).
2. Algebraic Geometry, Springer-Verlag, New York (1977).

Harvey, F.R.
1. Integral formulae connected by Dolbeault's isomorphism, Rice University Studies 56 (1970), 77-97.
2. Holomorphic chains and boundaries, Proc. A.M.S. Symp. in Pure Math., XXX, 2 (1977), 303-382.

Heins, M.
1. Selected Topics in the Classical Theory of Functions of a Complex Variable, Holt, Rinehart and Winston, New York (1962).

Hervé, M.
1. Several Complex Variables, Tata Inst., Bombay and OUP (1963).

Hille, B.
1. Analytic Function Theory 1, Ginn and Company, Boston (1959).

Hironaka, H.
1. Resolution of singularities of an algebraic variety over a field of characteristic zero, Ann. Math., 79 (1964), I: 109-203; II: 205-326.
2. Memorias de Matematica del Instituto "Jorge Juan", Madrid: 28, Introduction to the theory of infinitely near singular points (1974); 29, The theory of maximal contact (1975); 30, Desingularization theorems (1977).

Hirsch, M.W.
1. Differential Topology, Graduate texts in mathematics, Springer-Verlag (1976).

Hirzebruch, F. and Kodaira, K.
1. On complex projective spaces, J. Math. Pure Appl., 36 (1957), 201-216.

Hörmander, L.
1. An Introduction to Complex Analysis in Several Variables. D. Van Nostrand, Princeton (1967).
2. Linear Partial Differential Operators, Springer-Verlag, Berlin (1964).
3. L^2 estimates and existence theorems for the $\bar{\partial}$-operator, Acta. Math., 113 (1965), 89-152.

Hua, L-K.
1. Harmonic Analysis of Functions of Several Complex Variables in the Classical Domains, A.M.S. Translations of Math. Monographs, 6 (1963).

Husmoller, D.
1. Fibre Bundles, McGraw-Hill, New York (1966).

Jackson, D.
1. Note on rational functions of several complex variables, J. Reine Angew. Math., 146 (1916), 185-188.

Kobayashi, S. and Nomizu, K.
1. Foundations of Differential Geometry, Vol. 1, Wiley, Interscience, New York (1962).
2. Foundations of Differential Geometry, Vol. 2, Wiley, Interscience, New York (1969).

Kodaira, K.
1. On Kähler varieties of restricted type, Ann. Math., 60 (1954), 28-48.
2. On the structure of compact complex analytic surfaces, I, II, III, IV, Amer. J. Math., 86 (1964), 751-798; 88 (1966), 682-721; 90 (1968), 55-83, 1048-1066.

Kodaira, K. and Morrow, J.
1. Complex Manifolds, Holt, Rinehart and Winston, New York, (1971).

Kodaira, K. and Spencer, D.C.
1. On deformations of complex analytic structures II, Ann. of Math., 67, 3 (1958), 403-466.

Kohn, J.J. and Nirenberg, L.
1. A pseudoconvex domain not admitting a holomorphic support function, Math. Ann., 201 (1973), 265-268.

Lang, S.
1. Differential Manifolds, Addison-Wesley (1972).
2. Elliptic Functions, Addison-Wesley (1973).

Leray, J.
1. Le calcul différential et intégral sur une variété analytique complexe (Problème de Cauchy III), Bull. Soc. Math. France, 87 (1959), 81-180.

Levi, E.E.
1. Studii sui punti singolari essenziali delle funzioni analitiche di due o più variabili complesse, Ann. Math. Pura Appl., 17 (1910), 61-87.

Levinson, N.
1. Transformation of an analytic function of several variables to a canonical form, Duke Math., J., 28 (1961), 345-358.
2. A canonical form for an analytic function of several variables at a critical point, Bull. Amer. Math. Soc. (1960), 68-69.

Malgrange, B.
1. Ideals of differentiable functions, Tata Inst., Bombay (1966).

Mather, J.N.
1. Differentiable Invariants, Topology 16 (1977), 145-155.

Matsushima, Y.
1. Differentiable Manifolds, Dekker, New York (1972).

Milnor, J.W.
1. Lectures on Differential Topology, Princeton University Press, Princeton (1958).
2. Singular points of Complex Hypersurfaces, Ann. Math. Studies, 61, Princeton University Press, Princeton (1968).
3. On manifolds homeomorphic to the 7-sphere, Ann. Math., 64 (1956), 399-405.

Milnor, J.W. and Stasheff, J.D.
1. Characteristic Classes, Ann. Math. Studies, 76, Princeton University Press, Princeton (1974).

Moishezon, B.G.
1. On n-dimensional compact varieties with n-algebraically independent meromorphic functions, I, II, III, Amer. Math. Soc. translations, Ser. 2, 63 (1967), 51-177.

Mori, S.
1. Projective manifolds with ample tangent bundles, Ann. Math., 110 (1979), 593-606.

Mumford, D.
1. Algebraic Geometry I, Complex Projective Varieties, Springer-Verlag, Berlin, Heidelberg, New York (1970).

Nagata, M.
1. Local rings, Interscience, New York (1962).

Narasimhan, R.
1. Analysis on Real and Complex Manifolds, Masson and Cie, Paris, North-Holland, Amsterdam (1968).
2. Several Complex Variables, University of Chicago, Chicago (1971).
3. Introduction to the Theory of Analytic Spaces, Lecture Notes in Mathematics, 25, Springer-Verlag (1966).
4. On the homology groups of Stein spaces, Invent. Math., 2 (1967), 377-385.
5. Compact analytical varieties, L'Enseignment Math., 14(1) (1968), 75-98.
6. Imbedding of holomorphically complete complex spaces, Amer. J. Math., 82 (1960), 917-934.

Norgeut, F.
1. Sur les domaines d'holomorphie des fonctions uniformes de plusieurs variables complex (passage du local au global), Bull. Soc. Math. France, 82 (1954), 139-159.

Oka, K.
1. Domaines pseudoconvex, Tôhoku Math. J., 49 (1942), 15-52.
2. Domaines finis sans point critique intérior, Jap. J. Math., 23 (1953), 97-155.
3. Sur Les Fonctions Analytiques de Plusieurs Variables (collected papers of Oka on several complex variables), Iwanami Shoten, Tokyo (1961).

Pyatetskii-Shapiro, I.I.
1. Automorphic Function Theory and the Geometry of Classical Domains, Gordon and Breach, New York (1969).

Rauch, H.E.
1. A trancendental view of the space of algebraic Riemann surfaces, Bull. Amer. Math. Soc., 71 (1965), 1-39.

Remmert, R.
1. Holomorphe und meromorphe Abbildunden komplexer Raume, Math. Ann., 133 (1957), 328-360.

de Rham, G.
1. Varietés Differentiables, Hermann, Paris (1955).

Robert, A.
1. Elliptic Curves, Lecture notes in Mathematics, 326, Springer-Verlag (1973).

Serre, J.P.
1. Une proprieté topologoque des domaines de Runge, Proc. Amer. Math. Soc., 6 (1955), 133-134.

Siegel, C.L.
1. Topics in Complex Function Theory, Vol. 1, Wiely-Interscience, New York (1969).
2. Analytic Functions of Several Complex Variables, Lecture notes at the Institute for Advanced Study, Princeton, N.J., (1948) (reprinted with corrections (1962)).

Simha, R.R.
1. Holomorphic mappings between balls and polydiscs, Proc. Amer. Math. Soc., 6 (1966), 133-134.

Siu, Y-T.
1. Pseudoconvexivity and the problem of Levi, Bull. Amer. Math. Soc., 84(4) (1978), 481-512.

Siu, Y-T. and Yau, S.T.
1. Compact Kähler manifolds of positive bisectional curvature, Invent. Math., 59 (1980), 189-204.

Spivak, M.
1. Calculus on Manifolds, Benjamin/Cummings, Reading, Mass. (1965).

Springer, G.
1. Introduction to Riemann Surfaces, Addison-Wesley, Reading, Mass. (1957).

Sundararaman, D.
1. Deformations and classification of compact Complex Manifolds, Complex Analysis and its Applications, Vol. 3, IAEA, Vienna (1976), 133-180.

Swinnerton-Dyer, H.P.F.
1. Analytic Theory of Abelian Varieties, London Math. Soc. Lecture Notes, 14, Cambrdige Univ. Press (1974).

Teichmuller, O.
1. Bestimmung der extremalen quasikonformen Abbildungen bei geschlossenen orientierten Riemannschen Flächen, Preuss. Akad. Ber., 4 (1943).

Thimm, W.
1. Meromorphe Abbildungen Riemannschen Bereichen, Math. Z., 60 (1954), 435-457.

Tougeron, J.C.P.
1. Idéaux de Fonctions Différentiables, Springer-Verlga, Berlin, Heidelberg, New York (1972).

Ueno, K.
1. Classification theory of Algebraic Varieties and Compact Complex Spaces, Lecture notes in Math., 439, Springer-Verlag (1975).

Valentine, F.A.
1. Convex Sets, McGraw-Hill, New York (1964).

Vladimirov, V.S.
1. Methods of the Theory of Functions of Many Complex Variables, M.I.T. Press, Cambrdige, Mass. (1966).
2. The Laplace transform of tempered distributions, Global Analysis and its Applications, Vol. 3, IAEA, Vienna (1974), 243-270.

Waerden, B.L. van der
1. Modern Algebra, Vol. 1, Frederick Unger (1966).
2. Modern Algebra, Vol. 2, Frederick Unger (1964).

Walker, R.
1. Algebraic Curves, Springer-Verlag, New York (1978).

Weil, A.
1. Variétés Kählériennes, Hermann, Paris (1971).

Wells, R.O. and Wolf, J.A.
1. Poincaré series and the automorphic cohomology on flag domains, Ann. Math., 105 (1977), 397-448.

Whitney, H.
1. Complex Analytic Varieties, Addison-Wesley, Reading, Mass. (1972).
2. Local properties of analytic varieties, Differential and Combinatorial Topology, Princeton Univ. Press, Princeton (1965), 205-244.

Whittaker, E.T. and Watson, G.N.
1. A Course of Modern Analysis, 4th. Edition., Cambridge Univ. Press (1927).

Zariski, O. and Samuel, P.
1. Commutative Algebra, Graduate Texts in Math., Springer-Verlag, New York (1975).

Index